How to Chat With AI to Get Better Results

A Human-in-Charge AI Prompt Guide for Beginners to Fix Weak Answers, Protect Privacy, Save Time, and Use AI with Confidence

George Munson

GL DIGITAL PUBLISHING LLC

First Edition 2026

Contents

Introduction

Arthur Thomas, a retired newspaper editor, ran a writing workshop above a public library. On his desk rested a stack of blank paper beside a long-cold cup of tea.

He didn't seem like someone who would shape the digital age, but his old lessons help us understand today's technology-related challenges.

A student had walked up to his desk, frustrated and stuck. The student had been trying to write an article about a local craftsman who built wooden boats by hand.

"I sat with him for an hour," the student complained, dropping a blank notebook onto Arthur's desk. "He barely gave me anything. Just yes and no answers. He is a dry well."

Arthur looked up, his eyes sharp behind thick glasses. He picked up a red pen and tapped it against the desk. "There are no dry wells," he said softly. "There are only bad buckets. Show me the questions you asked him."

The student opened the notebook to show a list of three simple lines: What do you make? How long have you been doing this? Is it hard work?

Arthur sighed and shook his head. To keep readers engaged, emphasize asking open-ended questions that encourage storytelling and insights, helping them learn how to guide conversations effectively.

He took a fresh sheet of paper and wrote down three new questions. They did not ask for simple facts. Instead, they asked for stories, choices, and lessons learned.

"Go back," Arthur said. "Act like a partner in a conversation. Guide him, but let him do the heavy lifting. And when he gives you an answer, do not just write it down and walk away. Use his answer to ask the next, better question."

The next day, the student came back with pages full of stories and secrets from the craftsman. The well wasn't dry; the bucket just hadn't reached deep enough.

Today, millions of people are standing before a new kind of well. It is a digital one, housed inside a clean, white chat box on a computer or phone screen. We are told that this tool, artificial intelligence, can write our emails, organize our projects, teach us new languages, and save us hours of work every single day. By seeing AI as a helpful partner, you can feel more confident and less overwhelmed in your interactions.

Yet, using it can leave us staring at a blinking cursor, uncertain. Recognizing its limitations helps us avoid frustration and use AI more confidently, knowing it may need better prompts.

Instead, the screen fills with dry, boring text. It sounds like a robot pretending to care. It uses words no one would say in a conversation and gives advice that feels useless in real life.

We close the tab, feeling disappointed and thinking we're not tech-savvy. We worry about falling behind and feel overwhelmed by all the so-called magic prompts.

The truth is simpler and more encouraging. You don't need a computer science degree to use AI. You don't have to memorize secret phrases or learn to code.

Struggling with AI doesn't mean you're bad at technology. It just means you're treating a conversation partner like a search engine.

The Blank Box Trap

For the past twenty years, the internet has trained us to use a search box. We learned to type short, choppy word fragments into a search engine. We typed things like best chicken recipe, weather in Raleigh, NC, or how to fix a leaky sink.

The search engine did not talk back to us. It did not think, compose, or create. It was simply a digital filing cabinet. It looked through millions of existing folders, found the ones that matched our words, and dumped a list of links onto our screen. We did not need to have a relationship with the search box. We just needed to type the right keywords.

When we open an AI tool, the screen looks almost identical to that old search box. It is a single, clean bar waiting for our input.

This similarity can fool us. It makes us use old search habits on a new kind of tool.

When you type a search-style phrase into an AI, the tool gets confused. If you write help me with my project, the AI does not know who you are, what your project is, who will read it, or what you want the final result to look like. It is forced to guess.

Trying to help, it gives you a summary of common, generic information that doesn't connect to your real challenges.

There is another side to this trap. AI-generated answers often present information with polished grammar and a confident tone, making the response appear authoritative and leading people to see the first answer as definitive. If it's not right, we think the tool failed, or our prompt was wrong. We don't realize the first answer is just a draft we can improve.

You do not need to accept those dry, robotic drafts. You do not need to feel intimidated by the empty box.

If you change how you see the tool, you can be just as productive and comfortable as any tech expert. Remember, AI is a tool to empower you, not a threat. This perspective helps you feel more trusting and less intimidated when working with AI.

The Human-in-Charge Principle

This book is built on one belief: **You are the boss, and the AI is your partner**. Seeing yourself as a supervisor helps you feel more confident and in control of AI interactions.

We'll help you move from feeling anxious about tech to feeling confident as a supervisor. When you use AI, you're not dealing with a smarter mind. You're working with a skilled but sometimes forgetful junior assistant.

Think about how you would train a new assistant at your job or in your community. You would not point to a pile of messy files and say, "Fix this." You would not walk away and expect them to read your mind.

Instead, you would give them clear directions. You would tell them who the work is for, what tone to use, and what mistakes to avoid. You would review their first draft, point out any parts that do not sound right, and ask them to make specific changes.

Most importantly, you would never let them send an important letter or make a big decision without your approval. You would always keep your hand on the steering wheel.

This is the **Human-in-Charge Principle**. It divides the work of using AI into clear, manageable roles:

- **The AI assists; the human decides**. The AI can give you ten ideas for a project, but only you can choose the one that fits your team's real-world needs.

- **The AI suggests; the human verifies**. The AI can pull together facts and figures in seconds, but you must double-check the details to ensure they are true before sharing them.

- **The AI drafts; the human edits**. The AI can write a quick, clean first draft of an email, but you must shape the voice, add your personal touch, and make sure it sounds like a human being wrote it.

- **The AI organizes; the human understands**. The AI can group a long list of complaints into clean categories, but only you understand the emotions, politics, and relationships behind those issues.

With this mindset, the blank box feels less intimidating. A bad answer isn't the end; it's the start of a useful conversation.

Meeting Paula Jenkins

To see how this works in real life, let us visit a quiet home office in Raleigh, North Carolina.

It's 8:30 p.m. on a rainy Tuesday. The house is finally quiet. Paula Jenkins sits at her desk, the soft blue light from her laptop screen shining in her tired eyes.

Paula is 42 and works as a Project Operations Manager for a mid-sized company that creates training programs for businesses. She is highly organized, proud of her work, and naturally cautious. She is also incredibly tired.

Lately, her days feel like a race against endless messages and meetings, leaving her with scattered notes and replies.

In addition to her daily tasks, the company is preparing to roll out a major new training program for a client. Her manager sent a company-wide email urging everyone to use AI tools to improve efficiency and stay competitive.

Paula felt a cold spike of worry when she read that email. She did not want to look like she was falling behind.

Earlier this evening, she decided to try. She opened ChatGPT, stared at the blank white box for several minutes, and finally typed, "Help me with this project."

The screen flashed. Within seconds, the AI produced a long 800-word plan full of phrases such as 'synergistic alignment,' 'leveraging core competencies,' and 'maximizing stakeholder engagement.'

Paula read the first few paragraphs and felt tense. The plan was full of meaningless corporate jargon. It didn't address her client's needs, ignored her company's timeline, and sounded fake.

"I knew it," Paula muttered to herself, closing the laptop lid with a soft click. "This stuff is for programmers. I must not be cut out for this."

She rubbed her temples, feeling the usual frustration and exhaustion. She worried that if she didn't learn to use these tools, someone else might take her place. But every time she tried, it felt like a waste of time. She felt caught between a future she didn't understand and a present she could barely handle.

What Paula didn't know was that she was just one small change away from a big breakthrough. She didn't need to change who she was or become a tech expert. She just needed to learn how to talk to the assistant inside her computer.

The Journey Ahead

This book is a practical, plain-English guide to help you go from where Paula is now to feeling confident and in control.

We won't talk about complex computer science, coding, or the math behind AI. Instead, we'll focus on clear, repeatable ways to communicate that you can use every day, at work, in your writing, and while learning.

We will follow the stories of three real-world people who represent the different ways we encounter AI:

- **Paula Jenkins**, our busy project manager, who needs to tame her inbox, organize chaotic meetings, and manage complex tasks without losing her sanity or risking her company's private data.

- **Sam Vance**, a thirty-four-year-old small business owner and fitness creator in Nashville, Tennessee, who wants to write warm, engaging marketing copy and plan social media schedules without sounding like a robotic salesman.
- **Carol Higgins**, a sixty-seven-year-old retired schoolteacher in Sarasota, Florida, who wants to use AI safely to explore her hobbies, plan family trips, and decipher confusing medical and insurance paperwork.

By following their struggles and successes, you'll learn a simple, structured system you can start using right away.

First, we will explore the difference between a search box and a conversation partner. You will learn how to shift your mindset and build a natural dialogue with AI.

Next, you will discover the **Five-Part Prompt Formula**. This is a reliable, step-by-step blueprint that ensures you get specific, useful, and natural answers every single time you type a question.

You will also learn how to share the right amount of background detail with the AI while keeping your personal, financial, and company secrets completely safe. We will walk through clear boundaries for what you can safely paste into these tools and what must be scrubbed.

When the AI gives you a weak, robotic, or incorrect answer, you will not have to give up. You will learn the **Prompt Rescue System**, a seven-step troubleshooting guide for diagnosing and repairing bad answers on the fly.

We will also tackle the critical issue of safety. You will learn the Three-Level Trust Test to help you decide when to trust an AI's answer, when to double-check the facts, and when to consult a certified human professional.

Finally, we will show you how to build these methods into your daily routine. We will walk through practical workflows for work, writing, and learning, and we will guide you through a **7-Day AI Confidence Challenge** to help you build lasting, positive habits.

By the end of this book, you won't feel anxious when you see an empty chat box. You won't worry about being left behind by technology.

You'll stand confidently as the human in charge, using AI as a helpful partner to save time, lower stress, and strengthen your unique voice.

The blank box is waiting. Let's get started.

Chapter One

AI Is Not a Search Box

It Is a Conversation Partner

The morning of Tuesday, October 11, 2025, began like almost every other day in Sarasota, Florida.

By 7:15 a.m., the sun cast a gentle light over the palm trees. In her quiet kitchen, Carol Higgins followed her usual routine. She made coffee, measured out two scoops of decaf, and opened a can of food for Barnaby, her old orange tabby cat, who purred at her feet.

Carol was sixty-seven and had taught elementary school for almost forty years. She was naturally curious and organized, enjoying crossword puzzles, historical novels, and caring for her porch ferns. But this morning, she felt a quiet sense of worry at her kitchen table.

Next to her coffee mug sat a thick white envelope. It had arrived the day before, and seeing the return address made Carol tense up. It was from her insurance company.

Her Medicare supplement plan was raising its monthly rates, and her doctor's office was changing which insurance plans they would accept next year. Suddenly, her familiar medical routine felt confusing and uncertain.

Carol opened the envelope and found a thirty-page booklet printed in tiny gray letters. It was full of unfamiliar terms: deductibles, co-insurance, out-of-pocket maximums, Plan G, Plan N, formulary tiers, and prior authorization.

She tried to read the first page, but her eyes grew tired quickly. Frustration and defeat set in, just as they had before.

A few days earlier, her son had visited and installed a new app on her tablet. He told her it was a free AI assistant. "Just talk to it, Mom," he said, kissing her. "It knows almost everything. It's like having a helpful, smart assistant right next to you."

Carol picked up her tablet and opened the app. A blank white screen appeared, with a single gray box at the bottom. A thin black line blinked, almost inviting her to ask a question.

She suddenly felt unsure. What should she type? She decided to use it like the search boxes she used for recipes or checking the weather.

She tapped the screen and typed, "Tell me about Medicare."

She pressed send. The screen flashed, and within seconds, a huge block of text appeared. It included history, government dates, laws from the 1960s, and a confusing breakdown of Parts A, B, C, and D.

Carol scrolled through the long, dry response, feeling discouraged. It looked just like the confusing booklet on her table. The AI didn't know she was sixty-seven, lived in Florida, or just wanted to keep her regular doctor.

She closed the app and set the tablet on the counter. "This is just another search engine," she said to Barnaby, now asleep in a sunny spot on the floor. "It's just faster and louder. Maybe I'm not tech-savvy enough for these new tools."

Carol's disappointment is something millions of people feel every day. We hear about the new age of technology, but when we try to use these tools, we often end up feeling even more confused and alone.

She realized that the real issue wasn't her intelligence or tech skills. Many people feel overwhelmed by these tools, but understanding what's behind that blinking cursor can make a difference and help you feel more in control.

The Great Search Box Misconception

For more than twenty years, our brains have been trained by a single digital shape: the search box.

When the internet first became popular in the late 1990s, we had to learn how to find information. Early websites were organized like giant phone books. If you did not know the exact address of a website, you could not find it.

Then came the search engine. It simplified everything by giving us a single, empty box.

To use it, we had to learn a very specific, unnatural way of writing. The computer didn't understand full sentences. If you typed, I am feeling a bit cold today and I would love to know how to make a warm soup with the chicken I have in my fridge, the search engine would get confused by all the extra words.

So we learned to leave out the extra words and speak in "keywordese." We typed short, choppy phrases like "a chicken soup recipe."

The search engine used those keywords like a huge digital filing cabinet. It searched billions of web pages, found the ones with those exact words, and gave us a list of blue links. It didn't read the recipes or know if the soup was good. It just matched our words to what was already online.

Because of this habit, when we open an AI tool today, we see the empty box and blinking cursor and think, "This is a search box. I need to use short, choppy keywords."

This is the big misunderstanding of our technology era.

An AI tool isn't a search engine. It doesn't have a huge filing cabinet of web pages to search or blue links to give you. Recognizing this can help you feel more capable of engaging with AI as a language partner rather than a confusing search box.

Instead, it's a language network. It's built to process, understand, and create human communication. When you type a message, it doesn't look for matching files. It reads your words, figures out what you mean, and writes a new, custom response based on what it has learned.

When Carol typed "Tell me about Medicare," she treated the AI like Google and entered a broad, vague keyword.

Since the AI didn't know who Carol was or what she needed, it had to guess, which sometimes leads to overwhelming or inaccurate responses. Understanding that AI can improve with clearer communication and context can help users trust these tools and use them more effectively.

To get a helpful answer, Carol didn't need to learn special codes or buy complicated prompts. Instead, she could have started a conversation, which would have allowed her to receive more relevant, tailored information, highlighting the advantage of a conversational approach over traditional search methods.

The Shift From Links to Language

To see why this change is important, think about the difference between a map and a tour guide.

If you visit a historic city for the first time, a search engine is like a paper map. It shows every street, building, and park, giving you lots of information.

But the map can't talk to you. It doesn't know if you're tired, dislike museums, or want a quiet spot for tea. It just gives you all the options, and you have to figure out the rest.

An AI tool is meant to be like a patient, knowledgeable tour guide walking with you.

If you give the guide a map and say, "Tell me about this city," they might start listing facts such as when it was founded, its population, or its main industry. It would feel boring and dry.

But if you tell the guide, "I've always lived in a quiet town and get tired in crowds. I want to see beautiful old buildings for two hours without walking more than three blocks. Where should we start?"

The guide will adjust right away. They'll smile, lead you down a quiet street, show you a hidden garden, and share stories that match your interests.

This shows the power of language over links. A search engine gives you raw data, but an AI tool offers structure, understanding, and real communication.

Consider what happens when you ask both tools to help you with a messy task.

If you type how to organize meeting notes into a search engine, you will get ten articles written by different business experts. Each article will give you generic tips, like "use bullet points" or "write down action items." You still have to read all the articles, open a blank document, and organize your messy notes by hand.

If you give those same messy, disorganized notes to an AI tool and say, "Act as my assistant. Turn these scattered thoughts into a clean, professional summary with clear deadlines," the tool will actually do the work for you. It will read your specific words, understand who is responsible for what, and write a clear table of action steps.

It does not give you links to articles about organizing; it actually organizes your unique information.

The Conversation Partner Mindset

Once you realize that AI is a conversation partner, your entire approach to the empty box changes. You no longer have to worry about writing a perfect, flawless question on your very first try.

In a real conversation with a friend or a coworker, you do not expect them to read your mind. You do not stop talking to them if their very first sentence isn't exactly what you wanted to hear.

Instead, you build a bridge of understanding back and forth. You share details, you ask them to clarify what they meant, and you guide them toward the goal of the conversation.

This is the **Conversation Partner Mindset**. It relies on seven simple behaviors that you already use every single day when talking to other human beings:

1. **Goal Setting**: You begin by telling the other person exactly what you want to achieve.

2. **Context Sharing**: You give them the background story, explaining who you are, why this matters, and the limits you face.

3. **Format Specification**: You tell them how you want the information delivered, whether you want a quick summary, a detailed list, or a clean table.

4. **Correction**: When they make a mistake or misunderstand you, you gently point it out and ask them to try again.

5. **Follow-Up**: You do not walk away after their first response. You ask deeper questions to explore their ideas.

6. **Examples**: Ask them to use simple, real-world examples to explain complex ideas.

7. **Reasoning**: You ask them to explain why they suggest a certain path so you can understand their logic.

When you use these seven habits, the empty chat box no longer feels like a test. Instead, it becomes a place where you and your assistant work together to solve problems.

What AI Excels At (And Where It Struggles)

To be a good manager of your new digital assistant, you need to know which tasks they excel at.

Since AI can write well and use perfect grammar, it's easy to think it can do anything. But like any assistant, it has strengths and real weaknesses.

What AI Is Good At

- **Summarizing long, dense text**: You can paste a long article, a legal document, or a long email thread and ask the AI to give you the three most important points in plain language.

- **Brainstorming creative ideas**: If you are stuck trying to think of a title for a presentation, a gift for a friend, or ideas for a family dinner, the AI can generate dozens of creative options in seconds.

- **Adjusting tone and clarity**: You can write a rough, angry draft of an email and ask the AI to rewrite it to sound polite, professional, and calm.

- **Explaining complex concepts**: You can ask the AI to explain advanced topics, like inflation, blockchain technology, or photosynthesis, as if you were a ten-year-old beginner.

- **Creating structures and templates**: It can build clean checklists, step-by-step guides, weekly schedules, and comparison tables.

- Translating languages: It can translate text between different languages while maintaining the natural tone and warmth of the original message.

What AI Is Not Good At

- **Fact-checking its own data**: AI does not "know" facts the way humans do. It predicts words based on patterns. This means it can confidently invent fake dates, fake statistics, and fake book titles that sound completely real.
- **Accessing real-time or private information**: Unless you specifically paste the information into the chat, the AI does not know what happened in the news today, and it cannot see your private files, calendar, or emails.
- **Understanding human emotions and office politics**: The AI does not know that your coworker is sensitive about a certain project, or that your boss prefers brief updates because they are stressed.
- **Replacing certified experts**: AI should never be used as the final word for serious medical diagnoses, legal disputes, or major financial investments. It can help you prepare questions for your doctor, lawyer, or accountant, but it cannot replace them.

Output vs. Outcome

When most people use AI, they make the mistake of focusing entirely on the **output**.

The output is simply the words that the AI spits out onto the screen. If you ask for an email, the output is the email's text. If you ask for a recipe, the output is the list of ingredients and steps.

But a "Human-in-Charge" prompter focuses on the **outcome**.

The outcome is the real-world result you want to achieve once the conversation is over. It is the feeling, action, or solution you need in your real life.

Let us look at a simple example to see the difference.

Imagine you need to write a message to a neighbor whose dog barked all night long.

If you focus only on the output, you might type a quick prompt: Write a note to my neighbor about their barking dog.

The AI will generate an output. It might write a highly formal, slightly cold note that says: Dear Neighbor, I am writing to formally request that you address the noise disturbance caused by your canine companion during the late-night hours.

If you slip that note under your neighbor's door, what will the real-world outcome be? The neighbor will likely feel defensive, embarrassed, and angry. The relationship between you will become tense. The output was grammatically perfect, but the outcome was a disaster.

Now, imagine you focus on the outcome first. You want your neighbor to feel respected, you want to maintain a friendly relationship, but you also need to get some sleep.

With this outcome in mind, you talk to the AI like a conversation partner:

> Act as a thoughtful, friendly neighbor. I need to write a short note to my neighbor about their dog barking late last night. I want to sound warm, understanding, and collaborative, not angry or formal. We are usually very friendly, and I want to keep it that way. Keep the note under three sentences and suggest we talk about it over coffee.

The AI will generate a completely different response:

> Hi Sarah, I hope you're having a good week! Barnaby was extra vocal last night, and I had a bit of trouble sleeping, so I wanted to check in and make sure everything is okay with him. Let's grab coffee sometime soon to catch up!

By focusing on the outcome, you helped your digital assistant write something that protects your real relationships. You stayed in control of the emotional side.

Bad, Better, and Best Prompts

To help you see how this conversational shift works in practice, let us look at three different ways to ask an AI tool for help with a common, everyday task: planning a weekend trip.

The Bad Prompt (The Search Box Style)

> Plan a weekend trip to Savannah.

Why this fails: This prompt treats the AI like a Google search bar. It provides zero context, no details about who is traveling, no budget, and no specific interests. The AI will be forced to guess, returning a massive, generic list of popular tourist traps that you could have found on any basic travel website in two seconds.

The Better Prompt (The Basic Request)

> Write a three-day travel plan for a weekend trip to Savannah, Georgia. I am traveling with my husband, and we want to see historical sites and eat good food.

Why this is an improvement: This prompt is much better because it provides a clear goal (a three-day plan), a target audience (a husband and wife), and two general interests (history and food). The AI will return a decent, structured itinerary.

However, it still lacks the specific boundaries that make a trip truly enjoyable. It does not know their physical limits, their budget, or how they like to travel.

The Best Prompt (The Conversation Partner Style)

> Act as a knowledgeable, local travel guide who specializes in relaxed, historic trips. My husband and I are planning a three-day weekend trip to Savannah, Georgia, in early November.
> Here is our context: We are in our late sixties, prefer gentle walking tours to crowded bus rides, and want to avoid long stairs or steep hills. We love quiet, historic gardens and local Southern cooking, but we prefer quiet restaurants over noisy, trendy spots.
> Please give us the itinerary in a clean, day-by-day format. For each day, suggest one morning activity, one lunch spot, one afternoon rest period, and one dinner option.
> Before you finalize the plan, list three helpful questions I should ask our hotel to ensure our stay is comfortable and accessible.

Why this is the best: This prompt is incredibly powerful because it establishes a specific professional role (a local travel guide), shares rich human context (age, season of travel, physical preferences), sets strict limits (gentle walking, no steep stairs, quiet restaurants), defines a clear visual structure (day-by-day with specific slots), and includes a helpful quality check (questions to ask the hotel).

The result will be a plan that's truly customized, practical, and ready to use in real life.

Try It Now Exercise (5–10 Minutes)

The best way to get comfortable with AI is to try a successful conversation yourself. Open any free AI tool on your device and follow these three easy steps:

1. **The First Turn**: Think of a simple, everyday task you need to do soon (such as planning a simple family dinner, writing a polite

email to decline an invitation, or understanding a basic financial term like "compound interest").

2. **The Conversational Entry**: Instead of typing a short search phrase, copy and paste the following prompt into the box, filling in the bracketed details: > "Act as a friendly, patient [coach/advisor/assistant]. I am new to using AI, and I want your help with [insert your task here]. Before you give me any advice or write anything, please ask me three simple questions so you can understand my exact situation and give me a highly customized, useful response."

3. **The Dialogue**: Press send, read the AI's three questions, and type your answers back in plain, simple language. Watch how the AI adapts its very next response to match your exact needs.

Prompt Rescue Box

The Problem

The AI ignored your formatting requests, didn't pay attention to your personal context, or gave you a massive, overwhelming wall of dense text that is too tiring to read.

The Rescue Prompt

Copy this exact text into the chat box to reset the conversation:

> This response is a bit too long and detailed for what I need. Let us pause and simplify. Please take that exact information and rewrite it as a short, bulleted list using plain, conversational English. Put the most important action step in bold at the very top, and keep the whole response under 150 words.

Human-in-Charge Check

Before you use or share any text from an AI tool, take ten seconds to read it out loud.

As you read, ask yourself: Does this sound like something a real, caring person would actually say to someone else?

If a sentence feels stiff, too formal, or uses words you wouldn't say in real life, don't hesitate to change it. The AI is just your draft writer. You're the editor, the decision-maker, and the human in charge.

Carol's Medicare Resolution

Let us return to the quiet kitchen in Sarasota, Florida.

By 1:30 p.m. that Tuesday, the afternoon sun warmed the back porch. Carol sat at her kitchen table again, but the thick insurance booklet was closed. She held her tablet, looking more at ease.

After lunch, Carol decided to try the AI assistant again. This time, she didn't treat it like a search box. She remembered teaching children by asking guiding questions and chose to treat the AI like a new student teacher, eager to help but needing clear direction.

She opened the chat window, cleared her previous search, and typed a new message:

> Act as a patient, plain-English tutor. I am a retired schoolteacher and am trying to understand my Medicare supplement options for next year. I am feeling very overwhelmed by all the legal and financial terms in this booklet.
>
> Please explain the basic difference between Medicare Plan G and Medicare Plan N as if you were explaining it to a beginner. Use a simple, real-world analogy to show the difference in cost, and tell me what three questions I should ask my doctor's office to make sure they still take my coverage.

She tapped send and held her breath as the screen flickered.

This time, there was no overwhelming wall of gray text. Instead, the AI replied in a warm, encouraging tone:

> Hello! It is an honor to help a fellow teacher. Let's make this simple and stress-free.
> Think of Plan G and Plan N like two different ways to pay for a pass to an amusement park...

The AI went on to explain the two plans using a beautifully simple story about buying an "all-inclusive pass" versus a "discount pass where you pay a small fee at each ride." The language was clear, respectful, and completely free of confusing corporate jargon. At the very bottom, it listed three polite, direct questions, printed in clean bullet points, that she could read to her doctor's receptionist over the phone.

Carol felt a wave of relief. The tension in her neck faded. For the first time in weeks, she understood her options. She realized she wasn't being left behind by technology; she had just learned how to open the door.

She picked up her notebook and wrote down the three questions for her doctor's office. She felt organized, confident, and in control of her health decisions.

She tapped her screen one last time, typing, "Thank you." That was exactly what I needed.

The AI quickly replied, "You are very welcome!" I am so glad that it helped. Let me know if you want to explore any other part of the booklet together.

Carol smiled, set down her tablet, and poured a fresh cup of tea. She had stopped searching and started talking. In doing so, she took her first step as the human in charge of her digital future.

Am I Treating AI Like a Conversation Partner?

Use this quick checklist to review your interactions with AI and ensure you are getting the most useful, natural answers possible:

- ☐ **Did I state my ultimate real-world goal clearly?**
  (Did I explain what outcome I actually need to achieve once the conversation is over?)

- ☐ **Did I provide enough personal or professional context?** (Did I explain who I am, who my audience is, and what limits or rules I am facing?)
- ☐ **Did I ask for a specific, easy-to-read format?** (Did I request bullet points, a clean table, a short email draft, or a numbered checklist?)
- ☐ **Did I instruct the AI to use simple, plain English?** (Did I actively tell it to avoid corporate jargon, marketing hype, and robotic clichés?)
- ☐ **Did I ask a helpful follow-up question to improve the first draft?** (Did I treat the first response as a rough draft and ask the AI to polish, shorten, or adjust it?)
- ☐ **Did I read the final output out loud to verify its tone and accuracy?** (Did I make sure it sounds like a real human being before sharing it with others?)

By changing her approach from search box to conversation partner, Carol turned a moment of digital panic into a quiet victory. She realized she didn't need to know how the computer worked, how to guide a helpful conversation.

But as she looked at her neat handwritten notes, a new question came to mind.

It was great to get a clear answer this time. But how could she make sure she got this kind of practical result every time she used her computer, no matter how big or tricky the task? How could she avoid guessing altogether?

To do that, she needed more than a friendly attitude. She needed a reliable, step-by-step recipe, a simple five-part formula she could use to build a strong communication bridge for any task in her work, writing, learning, or daily life.

Chapter Two

The Five-Part Prompt Formula

If you walk down Tremont Street in downtown Boston today, you'll notice the city's fast pace all around you. Glass buildings tower overhead, and people in business suits rush by, focused on their phones and listening to music. Cars honk, delivery trucks roll over the streets, and the smell of coffee comes from a busy corner café.

But if you could go back in time, before the glass towers and paved streets, Boston would look very different.

In the fall of 1896, this area had cobblestone streets, coal smoke in the air, and the sound of horses' hooves on the road. Men wore wool coats and tall hats, while women walked in long skirts that picked up street dust.

A short walk from the main street was a drafty kitchen lit by gas lamps inside a brick building. This was the Boston Cooking School. The air was filled with the smells of rising yeast, roasting mutton, and wood vinegar.

In the middle of the kitchen stood a woman with a calm, determined look and a limp from childhood polio. Her name was Fannie Merritt Farmer.

On her wooden worktable was a stack of handwritten recipe books from that time. For Fannie, these books weren't helpful. Instead, they frustrated her.

In the late 1800s, people saw cooking as a kind of magic or secret talent, not a practical skill. If someone wanted to bake bread or roast a chicken, they had to use cookbooks with very vague instructions.

Recipes back then didn't use standard measurements. A cake recipe might say to use "a sweet pinch of salt," "a tea-cup of flour," or "a piece of butter the size of an egg."

But whose teacup did they mean? How big was the egg?

Even worse, the heat instructions were very unclear. Cookbooks said to use a "quick oven," a "slack oven," or a "gentle fire." Since stoves didn't have thermometers, cooks had to put their hand in the oven and count how many seconds they could stand the heat.

Because of this lack of structure, cooks across the country experienced kitchen disasters. The bread came out heavy and hard. Cakes sank into sticky middles. Roasts were either burnt on the outside and raw inside, or so dry they tasted like leather.

When meals turned out badly, young cooks didn't blame the unclear cookbooks. They blamed themselves. They sat in hot, smoky kitchens and cried, thinking they didn't have the "natural touch" or talent to be good cooks. They felt let down by a system that wouldn't give them clear instructions.

Fannie Farmer didn't accept this. She was logical and precise. She saw that cooking failed not because of the cooks, but because the instructions were unclear.

She believed anyone could cook a great meal if they had clear, repeatable steps.

Fannie got to work in the school kitchen, testing and measuring every ingredient. She stopped using vague teacups and pieces of butter the size of eggs.

Instead, she introduced something new to cooking: the level measuring cup and teaspoon. She said a cup of flour should be scraped flat with a knife so it's always the same amount. She also explained what a "quick oven" meant in real, measurable terms.

In 1896, she published the Boston Cooking-School Cook Book. It didn't have vague hints or flowery language. Instead, it had clear, structured recipes.

Her book was an instant hit. It sold millions of copies and changed how people cooked. She proved that cooking wasn't just for a few talented people. It was about clear instructions and steps anyone could follow.

Today, we're in a kitchen much like that one.

When we open an AI chat box, it's like getting an old, vague cookbook. We type in short, unclear requests and hope for the best. Asking the AI to "write a good email" or "give me some ideas" is just like asking for "a piece of butter the size of an egg."

When the AI gives us a dull or unhelpful answer, we react like the young cooks in Boston did. We close the screen, sigh, and think we just aren't "tech-savvy" enough to use these tools.

But the problem isn't your skill. The problem is the recipe. You don't need a special talent to get good results from AI. You need a reliable, five-part formula that shows you how applying a structured approach can consistently improve your responses.

Sam Vance's Nashville Struggle

To see what happens when we try to work without a formula, we must travel to a lively, growing neighborhood in East Nashville, Tennessee.

Inside a renovated brick warehouse, the air was filled with the rhythmic, heavy thud of medicine balls hitting a wall, the squeak of sneakers on black rubber mats, and the loud hum of a large industrial fan. This was Iron & Oak, a local gym owned by thirty-four-year-old Sam Vance.

Sam was a fit, energetic man with a warm smile and a neat beard. He had opened his gym three years ago with a simple mission: to help busy parents over forty get strong, move without pain, and find a welcoming community. He loved his work and his clients.

But on this hot Thursday afternoon, Sam was not on the gym floor. He was sitting at a small wooden desk in his cramped back office, staring at his laptop screen with a look of pure frustration.

The gym was doing, but when he sat down to write, his mind went completely blank.

He had heard his friends talk about how artificial intelligence could write months of social media posts in minutes. Hoping to save time and relieve his stress, Sam had opened a free AI tool on his computer.

He stared at the blank cursor for a moment, then typed: Give me content ideas.

The AI responded instantly, pouring out a list of twenty ideas. Sam read through them, his face falling:

- Post a photo of someone lifting weights with a motivational quote.
- Share a tip about drinking more water throughout the day.

- Write about the history of the barbell.
- Tell your followers to "crush their goals" on Monday morning.

The ideas were incredibly generic. They looked like every other boring fitness post on the internet.

Frustrated, Sam tried again, typing a slightly longer message: Write a Facebook post about working out.

The screen flashed, and the AI delivered a paragraph of text:

> Are you ready to unleash the beast within? It is time to stop making excuses and start crushing your goals! Embark on a life-changing fitness journey today and unlock your ultimate potential. Don't miss out on this incredible opportunity to transform your body and mind!

Sam cringed. He could never post that. It sounded like a loud, pushy late-night television commercial. It used words like 'embark,' 'unleash,' and 'ultimate potential, words he would never use to the real parents who walked into his gym.

His clients were not looking to "unleash the beast." They were looking to play with their kids without their lower backs hurting. They wanted a supportive, friendly space, not a high-pressure boot camp.

"This is useless," Sam muttered, rubbing his eyes. "It sounds like a plastic robot. If I post this, people will think I've lost my mind. I guess I'll have to spend five hours writing these posts myself."

Sam was stuck in the blank box trap. He had dropped a vague keyword into the AI, and the tool had returned a broad, average, and cliché-filled guess. He had no way of knowing that the AI was fully capable of writing warm, authentic, and highly effective posts in his exact voice.

He did not have the right measuring tools to shape the output.

The Core Formula

To get predictable, high-quality results from any artificial intelligence tool, you must stop guessing what to type.

Just as Fannie Farmer brought order to the kitchen with her level measurements, we can bring order to the chat box with a simple, five-part structure.

This is the **Five-Part Prompt Formula**. It is a repeatable blueprint that you can use for any task, whether you are drafting a professional email, organizing a project, learning a new language, or planning a trip.

The formula consists of five distinct pieces:

Goal + Context + Role + Format + Quality Check

When you combine these five elements into a single message, you remove the AI's need to guess. You set clear boundaries, provide rich background details, and maintain strict quality standards.

Let us break down each part of the formula to see how it works, why it matters, and how it transforms the AI's responses.

Part One: Goal (What do you want to accomplish?)

The **Goal** is the foundation of your prompt. It is the specific, primary task you want the AI to perform.

Many people make the mistake of using vague, passive language when stating their goal. They write things like "Help me with my newsletter" or "Tell me about marketing." These phrases don't give the AI a clear target.

To write an effective goal, you must use active, precise verbs. You must tell the AI exactly what physical action it needs to take.

- **Vague Goal**: "Help me write a gym email."

- **Precise Goal**: "Draft a three-paragraph invitation email for a free Saturday morning workout."

- **Vague Goal**: "I need to prepare for an interview."

- **Precise Goal**: "Generate five difficult practice interview questions for a project manager position."

By being specific about the action and the scope of the task, you instantly narrow the AI's focus. You prevent it from writing a giant, rambling essay when you only need a short, direct message.

Part Two: Context (What is the background?)

If the **Goal** is the skeleton of your prompt, the Context is the muscle and skin. It is the essential background information that shapes how the AI approaches the task.

Without context, the AI writes for the "average" internet user. But you do not live an average life, and your audience is not average.

To give the AI the right context, you should answer a few simple questions in your prompt:

- Who is this for? (Who is your target audience?)
- Who are you? (What is your background or business?)
- What are the main rules, limits, or struggles involved?
- Why does this task matter?

Let us look at how adding context changes the flavor of the AI's response.

If you ask for a "healthy meal plan" without context, the AI might suggest expensive ingredients, complex cooking techniques, and meals that take two hours to prepare.

But if you add context, "I am a busy single parent of two school-aged children, I have a budget of fifty dollars for the week, and I only have twenty minutes to cook dinner each night", the AI will completely change its suggestions. It will propose simple, budget-friendly, and quick recipes that fit your actual life.

In Sam's case, his context was missing. The AI did not know its target audience was parents over forty who felt intimidated by traditional gyms. It did not know that his gym's brand was warm, supportive, and focused on pain-free movement. Once he provides that context, the cheesy marketing clichés will disappear.

Part Three: Role (Who is the AI pretending to be?)

Assigning a **Role** is one of the most powerful ways to improve the quality of the AI's writing.

When you open an AI tool, it has access to a vast, sprawling library of language patterns. It can write like a lawyer, a pirate, a child, a scientist, or a pushy salesman. If you do not assign a role, the AI defaults to a neutral, slightly stiff "computer assistant" voice.

By telling the AI to adopt a specific professional perspective, you force it to select the right vocabulary, tone, and depth of knowledge for the task.

Think of it as hiring a specialized consultant for a few minutes. You can ask the AI to act as:

- A patient writing coach who helps simplify complex sentences.
- A direct-response copywriter who knows how to write engaging, authentic marketing copy.
- A meticulous project manager who excels at breaking down large tasks into clear steps.
- A plain-English medical translator who can explain complex billing codes.

When assigning a role, keep it realistic and task-focused. Avoid silly or exaggerated titles like "the world's greatest genius writer."

Instead, use practical roles that match the expertise you actually need.

- **Without a Role**: "Write a paragraph about stretching."
- **With a Role**: "Act as a physical therapist who specializes in gentle, daily mobility exercises for older adults."

The second prompt instantly guides the AI to use safe, encouraging, and anatomically accurate language.

Part Four: Format (How should the answer look?)

Many people get frustrated with AI because it gives them a massive, dense wall of text that is difficult to read and use. They waste precious time copying the text into a document and manually chopping it up into paragraphs, bullet points, or tables.

You can avoid this work entirely by specifying the exact **Format** you want in your prompt.

AI is incredibly skilled at organizing information visually. You do not have to accept a standard essay. You can ask for:

- A clean, two-column Markdown table (e.g., to compare options side by side).
- A chronological, step-by-step checklist.
- A series of short, three-sentence paragraphs.
- A bulleted list with the most important points printed in **bold**.
- A daily calendar structure.

If you are asking the AI to write an email or a social media post, specify the length and the parts you want included.

For example, you can write: "Please deliver this as a single email draft under 150 words. Include three distinct, engaging subject line options at the very top, and use short, one-sentence paragraphs to make it easy to read on a mobile phone." This level of specificity ensures the AI delivers exactly what you need on the first try.

Part Five: Quality Check (What are the boundaries?)

The final piece of the formula is the **Quality Check**. This is your defensive shield against bad, robotic, or inaccurate outputs.

Most people omit this step, which is why they end up with drafts packed with annoying clichés, long-winded introductions, and fake statistics.

A Quality Check is a specific instruction that forces the AI to review and correct its own draft before displaying it to you. It sets strict boundaries on what the AI must avoid or verify.

Here are some highly effective quality checks you can add to the end of your prompts:

- "Do not use corporate jargon or cheesy marketing clichés like 'delve,' 'unleash,' 'testament,' or 'in today's fast-paced world.'"
- "If you make any factual assumptions in this plan, list them clearly at the bottom."
- "Before finishing, check the reading level of the text and ensure it

is written at a simple, eighth-grade level."

- "If you do not have access to real-time local data for this request, state that clearly rather than guessing."

By setting these boundaries, you act as the firm supervisor, telling the assistant, "Here is the work I want, and here are the specific mistakes I do not want to see in your draft."

The Master Prompt

Now that we have explored all five parts of the formula, let us see how they fit together into a single, clean template.

You do not need to memorize this word-for-word, but keeping its structure in mind will completely remove the anxiety of the blank box:

> **Act as a [Role]. I want to [Goal]. Here is our context: [Context]. Please deliver the final output in this format: [Format]. Before you finish, perform this quality check: [Quality Check].**

This is the level measuring cup of the digital age. It is simple, repeatable, and completely under your control.

Sam's Social Media Breakthrough

Let us return to the cramped, warm back office of Iron & Oak in East Nashville.

It was now 4:15 p.m. The heavy thud of medicine balls had stopped, replaced by the soft, steady hum of the afternoon breeze through the open warehouse doors. Sam Vance was still at his desk, but his posture had changed. He was no longer slouched over, rubbing his eyes in defeat.

After reading about the Five-Part Prompt Formula, Sam decided to give his social media task one more try. He opened a fresh, clean chat window. He realized that his previous attempts had failed because he had given the AI zero measurements. He had asked for "butter the size of an egg."

This time, he took his time and built his prompt using the five-part recipe:

- **Role**: A direct-response copywriter who specializes in warm, welcoming, and non-intimidating fitness marketing for adults over forty.

- **Goal**: Draft three distinct social media post ideas to invite local parents to a free "Saturday Morning Strength" class.

- **Context**: The gym is Iron & Oak in East Nashville. The target audience is busy moms and dads over 40 who want to get stronger, reduce joint pain, and feel more energetic. They are often intimidated by loud, sweaty "hardcore" gyms. The tone must be friendly, encouraging, down-to-earth, and practical.

- **Format**: Three separate post options. For each option, include an engaging hook, a short body paragraph using one-sentence paragraphs, and a clear call to action.

- **Quality Check**: Do not use aggressive gym language, high-pressure sales tactics, or clichés such as "unleash," "no excuses," "crush your goals," or "embark on a journey."

Sam typed the completed prompt into the empty white box and pressed send. He watched the screen, his heart beating a little faster.

Within seconds, the AI began to write. This time, there was no loud, robotic salesman shouting from the screen. The text flowed naturally, sounding exactly like a conversation Sam would have with a friend at a local coffee shop:

Option 1: The "Playground" Post

Hook: Have you noticed that getting down on the floor to play with your kids, or grandkids, feels a little harder on your knees than it used to?
Body: You don't need to lift heavy tires or scream through intense workouts to stay active.
At Iron & Oak, we help busy parents over forty build real,

practical strength so they can move without pain.
Our coaches are patient, our community is friendly, and there are no loud mirrors or intimidating music.
Call to Action: Want to try a gentler way to get strong? Join us this Saturday at 9:00 a.m. for a free, beginner-friendly workout. Tap the link below to save your spot!

Sam read the screen, a slow, wide smile spreading across his face.
"Wow," he whispered to the empty room. "That is actually good. It sounds like me. It sounds like my gym."
The second and third options were just as strong. One focused on how to safely build core strength to protect a sore lower back during long office hours, and the other shared a lighthearted story about why rest is just as important as exercise. Both options were completely free of the aggressive clichés that had made Sam cringe earlier.
The difference was night and day. By using the Five-Part Prompt Formula, Sam had transformed the AI from a useless, generic text generator into a highly skilled, intuitive marketing assistant. He had kept his hand firmly on the steering wheel, resulting in a set of beautiful, authentic posts that he was proud to share with his community.

Try It Now Exercise (5–10 Minutes)

Now it is your turn to experience the power of a structured formula. Open your preferred AI tool, think of a simple task you need to complete today, and fill in the blanks of this template:

1. **Role**: Act as a [e.g., professional organizer, helpful neighbor, patient tech coach].

2. **Goal**: I want to [e.g., draft a polite email declining an invitation, create a packing list, write a recipe].

3. **Context**: Here is the situation: [e.g., I am a busy parent, the recipient is my boss, and I have a $20 budget].

4. **Format**: Deliver the response as [e.g., a bulleted list under 100 words, a clean two-column table].

5. **Quality Check**: Avoid using [e.g., formal legal language, corporate clichés, long introductions].

Run the prompt, read the output, and note how much more specific, natural, and usable the answer is compared to a simple, search-style question.

Prompt Rescue Box

The Problem

You used the Five-Part Prompt Formula, but the AI got confused by the amount of information and completely ignored your formatting request. It gave you a long, rambling narrative instead of the neat table or checklist you asked for.

The Rescue Prompt

Do not start a new chat. Type this exact message into the existing conversation:

> You ignored my formatting request and gave me a long paragraph. Please stop and restructure that exact information into the format I requested: a clean, two-column table. Place the main steps in the left column and the required tools or details in the right column. Keep it simple and direct.

This quick course-correction will force the AI to self-correct and deliver the clean, organized format you originally wanted.

Human-in-Charge Check

When assigning a professional "Role" to the AI, make sure the persona does not introduce unwanted bias, overly dramatic language, or a false sense of authority.

If you ask the AI to act as a "famous, dramatic chef," it might write a recipe that is fun to read but incredibly difficult to cook in a real kitchen.

Keep your assistant's role grounded, professional, and directly focused on the practical outcome you need. You are the manager; keep their focus on the work.

Personal Prompt Formula Card

Use this simple, physical card structure to sketch out your prompts in your mind, or on a scrap of paper, before you type them into the computer:

- **My Goal is**: __________
- **The Context is**: __________
- **The Role I want AI to play is**: __________
- **The Format I want is**: __________
- **The Quality Check I need is**: __________

Sam Vance stood up from his desk, stretched his arms over his head, and let out a long, happy breath. The heavy weight of writing anxiety that had hung over his shoulders for years was gone. He had a weekly schedule of social media posts ready to share, and he had completed the entire task in less than fifteen minutes.

He felt a deep sense of victory. He was no longer afraid of the blank white box. He knew exactly how to measure his ingredients to get a perfect result.

But as he reached for his mouse to copy the first social media post into his gym's scheduling tool, his hand paused.

To write the next batch of emails, Sam wanted to make them even more personal. He wanted to share a story about one of his real clients, a local mother who had struggled with chronic knee pain before finding strength at his gym. He wanted to paste some of her progress notes, her real goals, and her feedback directly into the prompt box so the AI could help him shape her story into an encouraging newsletter.

He opened the document containing his client's files. He highlighted the text, ready to copy and paste.

Suddenly, a quiet, cold chill of realization hit him.

The document contained her client's full name, age, personal email address, and detailed notes on her medical history and past injuries.

If he pasted this information into the free AI tool, where would it go? Who would see it? Could it be used to train the public model, allowing someone else to stumble across his client's private struggles?

Sam stared at the glowing blue screen, his finger hovering over the mouse button. The excitement of his breakthrough evaporated, replaced by a sudden, protective worry for the people who trusted him.

How much information was too much? Where did helpful context end, and dangerous oversharing begin?

Chapter Three

Give AI the Right Context Without Oversharing

"Gentlemen do not read each other's mail."

The words hung in the quiet office. Henry Stimson, the new Secretary of State, sat behind his large desk in Washington, D.C. It was spring 1929. Across from him, an assistant held a folder of typed papers.

The assistant stayed still. The only sound in the room was the slow ticking of a grandfather clock in the corner.

The folder held private messages from foreign governments. A secret group of analysts in New York City had intercepted, translated, and decoded them, illustrating how easily private communications can be accessed without permission, highlighting the importance of privacy awareness.

The Black Chamber's codebreakers were experts. At a major peace conference a few years before, they broke Japan's secret codes. They knew what Japan's leaders would accept before talks began, giving the United States a big advantage.

But when Stimson learned of the operation, he was upset and offended.

Stimson believed a gentleman's word was sacred. Reading private letters without permission to gain an advantage felt wrong to him. He thought a sealed letter belonged only to the person to whom it was addressed.

Stimson quickly stopped all government funding for the Black Chamber. He closed the office, locked up the codebooks, and sent the analysts home. He believed this would protect the privacy of human communication.

But Stimson's well-meaning action was based on a misunderstanding of how the world was changing.

He did not realize that the era of sealed paper letters was over. People now use the telegraph.

When a diplomat in Washington needed to send a message to Tokyo or London, they went to a telegraph office. They gave a sheet of paper to a clerk, who sent the message as clicks and buzzes over copper wires.

Diplomats thought their messages were private because they used complex codes. They trusted the telegraph's technology to keep them safe.

But they overlooked a simple fact: sending a message meant giving their words to someone else.

Telegraph companies kept copies of every message. Clerks read the words, operators heard the clicks, and government agents quietly collected the paper copies, illustrating how data can be accessed even when it seems private and underscoring the need for caution.

Diplomats were not keeping their secrets safe. They were giving them to strangers and hoping no one would read them.

Today, we are in a similar situation to those old telegraph offices.

When we use an AI tool, we see a clean, private screen. We type our thoughts, projects, and daily struggles from our offices or kitchen tables.

Since the screen is in our home and we are logged into our accounts, we feel safe, almost as if we are writing in a private diary.

The feeling that your prompt box is a private diary can be misleading, but understanding how to manage your data helps you maintain control over your privacy.

The prompt box is not a private diary. It connects your computer to a network of servers run by large companies.

When you paste your words into the box, they leave your computer. They are stored in data centers, analyzed by software, and sometimes used to train future AI models.

If you paste private company files, client details, or family medical histories into the box, you are sharing sensitive information online. Being aware of this helps you protect your secrets.

You do not have to stop using AI to protect your privacy. You do not need to avoid new technology.

Learning how to give your digital assistant enough context without oversharing allows you to use AI effectively while maintaining control over your data.

The Mind-Reading Myth

To understand why we make this mistake, we have to look at how our brains think about communication.

In our daily lives, we are surrounded by people who know us. Our spouses, our coworkers, and our close friends understand our history. They know where we work, who we are talking about when we mention a colleague, and what our daily struggles feel like. We do not have to explain everything from scratch every time we speak.

Because of this, we often fall into the **Mind-Reading Myth** when we use AI.

We assume the AI has a general understanding of our lives. We type a quick, brief sentence and expect the tool to fill in the blanks.

For example, a business owner might type: Write a polite email to a client explaining why their order is late.

This prompt lacks almost all the necessary context. The AI does not know:

- Who the client is.
- What they ordered.
- Why the order is actually late.
- How late will it be?
- What is your business's tone?

Because the AI is a language predictor, it cannot just sit there waiting for you to explain. It has to write something.

So, it falls back on its patterns. It generates a generic, stiff email that says: Dear Valued Customer, We regret to inform you that your shipment has been delayed due to unforeseen circumstances.

This response is useless. It sounds like a cold form letter from a giant bank.

To fix this, the business owner might try to solve the problem by dumping a massive pile of raw information into the box. They might copy and paste the entire customer order history, the shipping labels, the internal emails from the warehouse team, and even the customer's personal phone number and billing address.

They think: "If I give the AI all the data, it will write the perfect, custom email."

This is where the trouble begins.

In their hurry to provide context, they have overshared. They have sent private shipping tracking numbers, personal home addresses, and internal company discussions over the digital wire.

They have fallen into a trap. They believe they have only two choices: give the AI zero context and get a useless, generic answer, or give the AI all the context and risk exposing their private data.

There is a third, much better path. You can give the AI the exact, rich context it needs to write a beautiful, custom message while keeping your private secrets completely safe.

Paula's Copy-Paste Near-Miss

Let us return to the quiet office in Raleigh, North Carolina, where Paula Jenkins was working late on a rainy Monday evening.

The house was completely still. The only sound was the soft, steady hum of the refrigerator in the kitchen and the gentle patter of raindrops against the windowpane. Paula sat at her desk, her laptop screen illuminating her face.

She was working on one of the most difficult tasks of her week: writing a performance review for a junior project coordinator named Sarah Mitchell.

Sarah was a kind, hardworking employee who had been with the company for 2 years. But over the last six months, her work had slipped. She had missed three major deadlines on an important client account, and her communication with the team had become slow and spotty.

Paula knew the reason behind the slip. Sarah's mother had been diagnosed with a serious illness, and Sarah had been spending her weekends and evenings traveling back and forth to a hospital in another state. She was exhausted, stressed, and stretched to her absolute limit.

Paula wanted to write a review that was honest about the missed deadlines but also supportive, compassionate, and fair. She wanted to show that the company valued Sarah as a human being, while still setting clear professional goals for the coming quarter.

Writing this kind of review was challenging. Every word was important.

Paula spent an hour trying to write the draft by hand, but every sentence she typed sounded either too harsh or too soft. She felt stuck and frustrated.

She looked at the open AI chat window on her screen. After her success with the project plan last week, she felt a new sense of curiosity. Maybe her digital assistant could help her find the right words.

She opened the company's internal HR portal. She found Sarah's file. She highlighted the entire page, which included:

- Sarah Mitchell's full name and home address.
- Her annual salary is $58,500.
- Her Social Security number.
- Detailed dates of her sick leave and medical absences.
- Raw notes from a private conversation where Sarah had cried and shared her mother's diagnosis.

Paula pressed the keys to copy the text. She clicked over to the AI chat window, pasted the massive block of information into the prompt box, and began typing: "Rewrite this into a professional, supportive performance review."

Her finger hovered over the blue "Send" button.

Suddenly, she froze.

She looked at Sarah's name on the screen. She looked at her salary figures and her Social Security number. She looked at the raw, private notes about her mother's illness.

She thought about the company's strict privacy policy. She thought about the trust Sarah had placed in her when she cried in her office.

If she pressed that send button, where would that data go? Would it be saved on a server in another state? Could a programmer at the tech company see it? Could the AI use Sarah's personal details to answer questions for someone else in the future?

Paula suddenly felt worried. She slowly moved her hand away from the mouse.

"What am I doing?" she whispered to the quiet room. "This isn't my personal diary. This is Sarah's life."

With a trembling finger, she hit the backspace key, holding it down until the empty white prompt box was completely clear.

She felt relieved, but she was back at the beginning. She was tired, her neck hurt, and she still faced a blank screen and a tough draft.

She wanted the AI's help, but she refused to compromise her standards or risk her team's trust. She needed a way to share the situation without sharing the secrets.

The Safe Summary Method

The secret to solving Paula's dilemma is a simple, powerful technique called the **Safe Summary Method**.

This method helps you remove all private, identifying information from your prompt while keeping the main situation clear.

To understand how this works, think about how a doctor discusses a difficult case with a colleague.

If a doctor in New York wants to ask a specialist in California for advice on a rare condition, they do not send the patient's driver's license, name, and home address. Doing so would violate the patient's privacy and break the law.

Instead, the doctor strips the case down to its medical core. They say: "I am treating a 45-year-old female patient who is experiencing sudden joint pain and fatigue after a trip to New England."

The specialist doesn't need to know the patient's name is Martha or that she lives on Elm Street to give excellent medical advice. They only need the clinical context.

You can do the same thing with your AI prompts.

With the Safe Summary Method, you turn your private situation into a general, anonymous scenario. You take out names, numbers, and brands, but keep the relationships, emotions, and goals.

Let us look at how this changes a prompt from dangerous to safe.

The Dangerous Prompt (Overshared and Unsafe)

> Rewrite this performance review for Sarah Mitchell in our HR department. She makes $58,500 a year. She missed three major deadlines on the Smith account in June because her mother was in the hospital for cancer treatments in Charlotte. I need to tell her she has to improve by next month or we might have to let her go.

Why this is dangerous: This prompt exposes Sarah's full name, her exact salary, her department, her mother's medical condition, the specific city where she is receiving treatment, and the name of a real client account. If this data is leaked or used for training, Sarah's private life is exposed.

The Safe Prompt (The Safe Summary)

> Act as a compassionate, professional Human Resources director. Help me draft a performance review for a junior team member who is a valuable employee but has recently struggled with missed deadlines and slow communication. Here is our context: Over the past six months, this employee has faced a major, sudden family health crisis that required significant personal travel and emotional stress. We want to support them during this difficult time, but we also need to establish clear, professional expectations for their work goals over the next 30 days. Please write the draft in a warm, collaborative, and clear tone. Focus on creating a supportive improvement plan rather than sounding disciplinary. Keep it under 250 words.

Why this is safe and highly effective: This prompt contains zero private data. There are no names, no salaries, no specific illnesses, and no client names.

Yet, the emotional and professional context is incredibly rich. The AI knows exactly what the relationship is, what the struggle is, and what the desired outcome is.

The AI will create a balanced, respectful draft that Paula can copy into her document and personalize with Sarah's real name on her own computer.

What to Share vs. What to Protect

To help you with this in daily life, you need a simple guide. You should know which files and details are safe to share with an AI tool, and which ones you must protect.

What Never to Share

Think of the prompt box as a public space. If you do not want to share the information on a big sign in your town, do not paste it into a public AI tool.

Keep these items completely out of your prompts:

- **Personal Identification**: Social Security numbers, driver's license numbers, passport details, and birth dates.
- **Financial Details**: Credit card numbers, personal bank accounts, salary figures, tax documents, and corporate financial ledgers.
- **Medical Records**: Specific diagnoses, patient names, mental health histories, and prescription details.
- **Security Data**: Account passwords, PINs, security questions, and server access keys.
- **Company Secrets**: Proprietary software code, unannounced product plans, legal strategies, and private corporate contracts.
- **Real Names**: The actual names of your clients, employees,

children, or colleagues.

What Is Safe to Share

You can share any context that explains your problem in general terms, as long as you do not reveal specific identities.

These items are safe and highly helpful for the AI:

- **General Roles**: "I am a project manager," "I am a retired teacher," or "I am a small business owner."
- **General Audiences**: "My target reader is a busy parent over forty," or "This is for a senior executive who prefers brief updates."
- **Tones and Emotions**: "I want to sound warm and collaborative," or "The tone should be direct, polite, and firm."
- **Anonymized Situations**: Using placeholders like "Employee A," "Client B," or "Company X."
- **Public Information**: Any text from public articles, book chapters, or general historical facts.

Few-Shot Prompting with Safe Examples

One of the most effective ways to get excellent writing from an AI is to show it examples of what you want. This is a technique called **Few-Shot Prompting**.

Instead of just describing your style, you paste a few sentences of your past writing into the prompt and say, "Write the new draft in this exact style."

This works incredibly well, and in private discussions, you send that data over the wire.

To use this technique safely, you must actively "anonymize" your examples before pasting them.

Look at this simple process for prepping your writing samples:

1. Copy the past email or document you want to use as a style

sample.

2. Paste it into a temporary document on your own computer (such as Notepad or Word).

3. Read through the text and replace all real names with placeholders like Diana or [Project X].

4. Remove any specific financial figures, dates, or locations.

5. Once the sample is clean, paste it into the AI prompt with a clear instruction: "Use this anonymized sample to understand my natural writing style, then write a new message about [Situation] using the same tone."

By following these steps, you get the benefits of customized writing without sharing your real relationships.

Try It Now (5-10 Minutes)

Let's practice the Safe Summary Method right now so you can see how easy it is to protect your privacy and still get excellent results.

1. **Think of a Real-World Task**: Think of a recent, slightly difficult email or message you had to write. It could be a polite message to your boss about a project delay, a note to a utility company disputing a charge, or a message to a family member about a sensitive holiday plan.

2. **Scrub the Details**: On a piece of paper or in a blank document, write down the core situation, but actively replace all real names, specific money figures, and private details with general terms.

3. **Run the Safe Prompt**: Open your AI tool and paste the following template, filling in your scrubbed details: > "Act as a [Role]. Help me write a [Format] about [Anonymized Situation]. My goal is [Goal]. The tone should be [Tone]. Please do not use any real-world names, specific financial numbers, or private identifying details. Keep the final response under 150 words."

4. **Compare the Results**: Read the generated draft. Notice how the AI delivered a highly specific, emotionally accurate response without knowing a single private detail about your life.

Prompt Rescue Box

The Problem

You were in a hurry and pasted a block of text into the AI, only to realize a second later that you forgot to remove a colleague's real name, a private project code, or a personal email address.

The Rescue Steps

Do not panic, but act immediately to secure your data:

1. **Do Not Continue the Conversation**: Stop typing in that specific chat window. Do not press send on any further messages in that session.

2. **Clear the Session**: Most AI tools have an option to delete the current chat history. Click the "Delete Chat" or "Clear Conversation" button immediately to remove it from your active dashboard.

3. **Reset in a Fresh Window**: Open a completely new, blank chat window.

4. **Run the Reset Prompt**: Paste this exact message to establish a firm boundary for the new session: > "We are starting a new, secure session. For this and all future requests, I will use generic placeholders such as 'Employee A' or 'Company X.' Do not attempt to guess or recall any real-world names, personal details, or specific metrics from previous sessions. Keep all responses strictly general and focused only on the structural context I provide."

Human-in-Charge Check

Always treat the prompt box as if it were public.

Before you click the send button on any prompt, take three seconds to look at the text box. Ask yourself this simple, powerful question:

"If this text were printed on a big sign in a busy place, would I feel comfortable and safe for anyone to read it?"

If the answer is no, take a moment to delete the private details and apply the Safe Summary Method. You are the boss; you are responsible for keeping your team's and your family's data safe.

Paula's Safe Resolution

Let us return one last time to the quiet home office in Raleigh, North Carolina.

The grandfather clock in the living room chimed 9:30 p.m. The rain had slowed to a gentle, quiet mist against the window glass. Paula Jenkins sat back in her chair, a soft, satisfied smile on her face.

The screen in front of her did not contain a giant, scary block of Sarah Mitchell's private HR files. Instead, the AI chat window showed the safe, anonymized prompt Paula had typed using the Safe Summary Method.

Directly below her prompt, the AI had delivered a beautiful, warm, and highly professional draft:

> I want to take a moment to recognize the dedication and care you bring to our team every day. Your presence is highly valued here.
> Over the last few months, we understand that you have faced significant personal challenges. We want to support you through this difficult season, and our primary goal is to help you succeed.
> To ensure we stay on track together, let's focus on a simple plan for the next 30 days. We want to help you rebuild your regular communication rhythm with the team and establish manageable timelines for our current accounts...

Paula read the words out loud. Her voice sounded soft and steady in the quiet room.

The tone was absolutely perfect. It did not sound robotic, cold, or clinical. It sounded like a caring, professional manager who wanted to help a struggling employee find her footing.

And best of all, there was not a single mention of Sarah's salary, her mother's specific diagnosis, or the cities she had been traveling to. Sarah's private life remained completely safe, locked inside the secure HR database on Paula's local computer.

Paula highlighted the text on her screen, copied it, and pasted it into her company's official performance review document.

Once the text was safely inside her own secure, private software, she manually typed Sarah's real name at the top of the page. She added a few specific, personal details about their local office schedule and saved the file.

She felt proud of what she had done. She did not let fear of technology or a busy schedule lower her standards for privacy and trust. She used AI as a helpful partner, but stayed in control. She was clearly the one in charge.

She closed her laptop lid with a soft, satisfying click. She stood up, stretched her back, and walked over to turn off the desk lamp.

But as she walked out of the room, her mind began to drift toward the next morning.

The draft she had created with the AI was a wonderful start. But as she thought about her upcoming meeting with Sarah, she realized that a simple draft was only the first step.

What if Sarah reacted with worry? What if the first draft of the performance review was a little too soft, and she needed to make it slightly more structured without starting from scratch?

How could she continue the conversation with her digital assistant to shape, polish, and refine the wording until it was absolutely flawless? How could she avoid the trap of starting a brand-new prompt every time she wanted to make a minor change?

To do that, she would need to learn new conversational skills: the art of the follow-up loop, where AI communication becomes truly effective.

Chapter Four

The Follow-Up Skill

How to Turn a Weak Answer Into a Useful One

The astronauts had just thirty-six hours before the air in their small spacecraft would become deadly.

On the evening of April 15, 1970, three astronauts were hurtling through the dark, cold vacuum of space, more than 200,000 miles from Earth. An explosion had crippled the Apollo 13 spacecraft. To survive, Jim Lovell, Jack Swigert, and Fred Haise had to crowd into the Lunar Module, a small, fragile vehicle designed to hold only two people for a short stay on the moon.

Now, it was their only lifeboat, but the air inside was quickly running out.

Every time the three men exhaled, they released carbon dioxide into the cabin. Under normal conditions, specialized canisters filled with chemical pellets scrubbed this invisible gas from the air. But the Lunar Module was never meant to support three grown men for a four-day journey back to Earth.

The warning lights on the instrument panel began to glow a dangerous, steady yellow. The level of carbon dioxide in the air was rising. If it went too high, the astronauts would first become confused, then suffer severe headaches, and eventually slip into unconsciousness and suffocate.

The spacecraft did have plenty of spare canisters. They were stored in the dead Command Module at the end of the tunnel. But there was a terrible, mechanical catch.

The canisters in the Command Module were square. The openings in the Lunar Module's filtration system were round.

A square peg will not fit into a round hole.

Down on Earth, inside a bright, windowless room at the Manned Spacecraft Center in Houston, an engineer named Ed Smylie was handed a mission that seemed impossible. He had to find a way to connect those square canisters to the round holes using only the items the astronauts had on board the crippled spacecraft.

He could not send a rescue ship with new parts. He could not mail them a tool. He had to work with what was already there: plastic bags, cardboard covers ripped from flight manuals, gray duct tape, and the flexible hoses from the astronauts' space suits.

Smylie and his team sat at a wooden table, surrounded by the same random items. They did not aim for a perfect machine on their first try because they did not have time.

Instead, they began with a rough idea. They taped a plastic bag to a square canister, but it leaked. They added more tape. When the hose did not fit well, they filled the gap with cardboard.

They tested, failed, made adjustments, and tried again, repeating this process over and over.

After building a rough working model in Houston, the real challenge started: explaining how to assemble it to the tired astronauts over a crackling radio connection.

This was not a simple broadcast. Smylie could not just read instructions and end the call. The astronauts were shivering and working in near-freezing temperatures with little sleep.

Giving instructions turned into a careful, ongoing conversation.

"Okay, Jack," the communicator in Houston said over the radio. "We want you to take a plastic storage bag. Do you have one of those handy?"

"We have the bag," Jack Swigert replied from space. "But it has a small tear near the seam. Will that be a problem?"

The engineers in Houston huddled around their table. They looked at their own bag. "Yes, that will leak," they realized.

Houston came back on the air: "Okay, skip that bag. Use the thicker plastic bag from the food storage locker instead. Slide the canister inside it, and let us know when you have done that."

"Okay, it is inside," Swigert said a few minutes later. "Now, what do we do with the manual cover?"

"Cut the cover about four inches from the top," Houston guided. "But wait, make sure you do not cut the binding, or the cardboard will fall apart."

"Understood," Swigert said. "Cutting it now. Wait, the scissors are slipping on this thick cardboard. Can we use a knife instead?"

"A knife is fine," Houston responded. "Just go slow so you do not puncture the plastic bag underneath."

The radio conversation continued for hours, showing a perfect example of a follow-up loop.

Every time Houston gave an instruction, the astronauts checked their physical reality, reported back what was working or failing, and asked for adjustments. Every time the astronauts encountered an obstacle, the engineers on the ground changed their plans, refined their measurements, and offered a new option.

They did not succeed on the first try. They succeeded because they kept communicating until the job was finished.

By the end of the long, tense night, the astronauts had built a strange box wrapped in gray tape and plastic. It looked like trash, but when they connected the hose, they heard air rushing into the cabin.

Within minutes, the carbon dioxide levels dropped, and the air became fresh and clean again.

The three men survived not because NASA had a ready-made solution, but because they used ongoing conversation to turn scraps into a lifesaver.

Carol's Packing List Crisis

To see how this same conversational loop works in our everyday lives, let us return to Sarasota, Florida.

It was 9:15 p.m. on a Tuesday. Carol Higgins was standing in her bedroom, staring at her bed with a look of quiet panic.

Her flight to Denver, Colorado, was scheduled to leave at 7:00 a.m. the next morning. She was traveling to visit her daughter and her young grandchildren for a five-day autumn trip.

Packing had always stressed Carol out. She was used to warm weather, sand, and light cotton shirts. The thought of going somewhere with falling leaves and sudden cold felt as strange to her as going to the moon.

Her bed was covered with clothes. She had packed three winter coats, two pairs of wool boots, four heavy books, three sweaters, two swimsuits, and a beach towel. Her suitcase was so full that the zipper would not close.

"I can't even lift this bag," Carol said, rubbing her sore back. Barnaby, her orange cat, watched her with sleepy eyes. "I'm sixty-seven. I can't drag a fifty-pound suitcase through two airports. I'll hurt my back before I even see the kids."

She remembered her success with the AI assistant from the previous week. She picked up her tablet, opened the chat window, and asked for help.

She typed, "Give me a packing list for Colorado."

The screen flashed. Within seconds, the AI delivered a massive, thirty-item list. It told her to pack:

- Heavy winter parkas and snow pants.
- Thermal underwear and wool socks.
- Hiking boots and trail gaiters.
- A water filtration bottle and a compass.
- Sunscreen, a sun hat, and a swimsuit.

Carol stared at the screen, feeling anxious. The list was so long that she would need two big suitcases. It did not consider her age, the short trip, or that she was staying in a warm house, not camping on a mountain.

For a moment, Carol wanted to close the app and give up. She felt the AI had let her down with an unrealistic, overwhelming answer.

But then she remembered the lesson about being in charge of the conversation. The AI's first answer was not final. It was just a rough draft, like a pile of virtual scraps she could shape.

She tapped her finger on the screen and began talking back to her assistant.

"This list is far too long," Carol typed. "I am sixty-seven years old, and I want to travel with only one light, carry-on suitcase. I am staying at my daughter's house in a Denver suburb for just five days in mid-October. We are going to walk in local parks and go to backyard dinners, but I am not hiking up any mountains. Please make this list simple, and focus on items I can layer."

The AI responded instantly, changing its tone to a warm, practical helper:

"Of course! Let's simplify this so you can travel light and stay comfortable. October in Denver is beautiful, but the weather can change quickly, from sunny afternoons to chilly evenings. The secret is layering.

Here is a simplified, five-day packing list that will easily fit into a single carry-on bag..."

The new list was much shorter. It suggested packing three light, long-sleeve shirts, two pairs of comfortable pants, one warm cardigan, and one medium jacket to save space on the plane. It removed the heavy snow pants, the compass, and the hiking boots entirely.

Carol felt her shoulders relax. "This is much better," she thought.

But as she looked at her bed, she realized she still had a question. "What should I actually wear tomorrow morning when I get on the plane?" she wondered. "Florida will be eighty degrees when I leave, but Denver will be forty degrees when I land."

Instead of starting a new search, she typed a follow-up:

"Thank you. That is very helpful. Now, tell me exactly what I should wear on the plane tomorrow morning so that I am comfortable in the warm Florida airport, but ready for the cold weather when I arrive in Denver. Keep in mind that I want to wear my heaviest jacket on the flight, so it does not take up space in my bag."

The AI replied with a neat, step-by-step outfit plan:

"Here is your perfect travel outfit:

1. **Base Layer**: Wear your lightest cotton t-shirt and your most comfortable walking sneakers. This will keep you cool as you walk through the airport in Florida.

2. **Middle Layer**: Carry a soft cardigan or sweater in your hands or drape it over your shoulders.

3. **Outer Layer**: Wear your medium jacket on the plane. If you get warm during the flight, you can easily slide it off and place it in the overhead bin.

4. **Bottoms**: Wear your most comfortable, stretchy pants for the flight.

This way, you save maximum space in your carry-on, and you can easily add or remove layers as the temperature changes."

Carol smiled. It was so simple, yet she had not thought of it.

Within ten minutes, she had sorted through her pile of clothes. She folded the selected layers into her small suitcase. The zipper closed with a smooth, effortless slide. The bag was light, neat, and easy to lift.

She turned her travel anxiety into a calm, organized win. She did this not by finding a perfect prompt, but by staying in the conversation and guiding her assistant until the answer fit her needs.

Why the First Answer Is Usually a Draft

When we use an artificial intelligence tool, we must understand a fundamental truth about how these systems are built: **The AI is programmed to give you the most average answer first.**

This is not because the AI is stupid or lazy. It is because of how it learns.

AI models are trained on billions of pages of publicly available text written by millions of people. When you ask a broad question like "How do I plan a garden?" or "Write a business email," the AI doesn't know your personal style, budget, or exact goals.

AI models are trained on billions of pages of publicly available text written by millions of people. When you ask a broad question like "How do I plan a garden?" or "Write a business email," the AI doesn't know your personal style, budget, or exact goals.

This is what we call **The Baseline Trap**.

If you accept this first, average response, you are letting the machine decide the quality of your work. You are accepting a draft that is dry, generic, and completely lacking in personal warmth or specific detail.

To become a Human-in-Charge prompter, you need to change your expectations. Treat the AI's first response as a starting point, like a lump of clay you can shape.

Your job is not to look at the clay. You need to work with it and shape it into what you actually need.

This shaping process happens in the **Editing Loop**.

Instead of typing a prompt, reading a disappointing answer, and starting over, learn to stay in the same chat window. Treat the AI's first draft as the beginning of a more detailed conversation.

With focused follow-up questions, you can help the AI fix mistakes, add details, remove extra words, and improve the tone until the result sounds natural.

The Five Best Follow-Up Questions

To guide your digital assistant effectively, you do not need to invent new questions from scratch every time.

Just like Ed Smylie and his team of engineers in Houston, you can rely on a set of reliable, highly effective communication tools.

Here are the five best follow-up questions you can use to rescue a weak, generic, or overwhelming response and turn it into a high-quality win:

1. The Clarifier

> Before you answer, ask me three clarifying questions to make your response more accurate.

When to use it: Use this when you are starting a complex task, such as writing a business plan or organizing a major family event, and you are not sure what details the AI needs to know to give you a good answer. This shifts the burden of extracting context onto the assistant.

2. The Concrete Example

> Make this response more specific and replace the general advice with three concrete, real-world examples.

When to use it: Use this when the AI gives you dry, academic, or theoretical advice. If it tells you to "improve customer communication," this prompt forces it to show you exactly what that looks like in daily practice.

3. The Jargon Cutter

> Make this response 50% shorter, remove the corporate jargon, and focus on clear, active steps.

When to use it: Use this when the AI delivers a massive, boring wall of text filled with buzzwords like synergy, leverage, or robust framework. This forces the AI to simplify its language and get straight to the point.

4. The Blind Spot Finder

> What key perspective, risk, or alternative option am I missing in this plan?

When to use it: Use this when you are making a decision, planning a project, or writing an argument. It forces the AI to act as a critical thinking partner, highlighting risks or ideas you might have overlooked in your excitement.

5. The Style Calibrator

> Give me three different style options for this draft, one warm and conversational, one brief and direct, and one formal, and explain when to use each one.

When to use it: Use this when you are writing an email, a letter, or a blog post, and you are not sure what tone will get the best response from your reader. This allows you to compare different emotional paths before making your final decision.

How to Correct AI Politely and Clearly

Many people feel strange correcting an artificial intelligence. Because the tool answers with such high confidence, we feel like we are being rude or that we are the ones who made a mistake.

Some people even type "please" and "thank you" in every prompt, worrying that they might hurt the computer's feelings.

You do not need to worry about the AI's feelings since it is just a program, but being clear, respectful, and structured in your corrections is still very helpful.

When you correct an AI, you give it new patterns to learn from. If you type, "That was terrible, try again," the AI does not learn why the draft was bad or what to change.

To effectively correct your assistant, use the **Reconstruction Prompt**.

Use this simple, three-part feedback structure:

1. **Praise**: Point out what the AI did well.

2. **Cut**: Say what needs to be removed or changed.

3. **Guide**: Explain what rule or tone to use for the next draft.

Let us look at a real-world example of how this works.

Imagine the AI drafted an email to your landlord requesting a repair, but it came across as too aggressive and angry.

Instead of starting over, you type:

> The explanation of the leaky faucet is excellent and accurate. However, the tone comes across as too demanding and confrontational. Please rewrite this draft to sound polite, cooperative, and respectful. Keep the specific details about the leak, but make sure we maintain a friendly relationship with the landlord.

This clear feedback helps the AI keep the good parts and fix the tone, saving you time and letting you stay in control.

How to Make AI Ask You Questions

One of the most underused secrets of great prompting is turning the tables on the AI.

Most people assume they have to do all the heavy lifting of thinking, organizing, and writing the perfect prompt. They sit at their desks, sweating over the keyboard, trying to remember every single detail.

You can let your digital assistant do that work for you by using the **Interviewer Prompt**.

If you are tackling a task where you feel out of your depth, such as writing a resume, planning a budget, or preparing for a difficult conversation, you can instruct the AI to interview you.

Here is how to run this tactic:

> Act as an expert career coach. I want to write a professional cover letter for a library assistant position. Instead of writing it now, please act as an interviewer.
> Ask me up to five questions, one at a time, about my experience, my values, and why I want this job. Once I have answered all five questions, I will use my answers to write a custom, warm cover letter in my voice.

This is incredibly powerful. Instead of staring at a blank page, you answer five simple, conversational questions.

The AI organizes your thoughts into a clear, professional structure. You stay in charge by simply answering its questions.

How to Ask for Options Instead of One Answer

When we face complex human problems, there is rarely one "perfect" solution.

If you are trying to resolve a conflict with a coworker, plan a career change, or decide how to handle a family budget, accepting a single piece of advice from an AI can be dangerous. It traps you in a single way of thinking.

To avoid this, you should use the **Spectrum Prompt**.

This prompt makes the AI give you several clear options. It prevents the AI from giving a single average answer and lets you compare different risks, efforts, and tones.

Here is an example of a Spectrum Prompt:

> I need to talk to my manager about feeling overwhelmed by my current workload. Please give me three distinct strategies for how to approach this conversation:
>
> - **Option A (The Conservative Path)**: A gentle, collaborative approach that focuses on prioritizing my current tasks.
> - **Option B (The Balanced Path)**: A direct, professional approach that proposes a specific shift in my project responsibilities.
> - **Option C (The Bold Path)**: A firm approach that establishes clear boundaries and requests immediate, temporary support.
>
> For each option, write a short, three-sentence script showing how I can start the meeting.

By looking at this spectrum, you can select the exact path that matches your unique relationship with your manager, your office politics, and your personal comfort level.

You are not letting the AI decide for you. Instead, you use its suggestions to explore your options before acting.

Bad, Better, and Best Follow-Up Examples

To see how these follow-up loops transform raw drafts into useful results, let us look at three different ways a professional might try to adjust a workplace email.

The Bad Follow-Up (The Vague Command)

> Make it better.

Why this fails: This follow-up gives the AI zero direction. It does not explain what was wrong with the first draft. Was it too long? Too formal? Too aggressive?

The AI will shuffle the words around, rewrite the sentences with different grammar, and hand back another generic draft that is likely no better than the first.

The Better Follow-Up (The Basic Adjustment)

> Rewrite this email to make it sound friendlier and shorter.

Why this is an improvement: This follow-up is much better because it provides two clear directions: friendlier tone and shorter length. The AI will make the email more accessible and trim some unnecessary sentences.

However, it still leaves a lot to chance. It does not specify the target length or define what "friendly" means in this specific professional relationship.

The Best Follow-Up (The Collaborative Edit)

> This draft is a good start, but it sounds a bit defensive. Let's refine it.
> Keep the second paragraph explaining the technical delay on the project.
> However, rewrite the opening and closing sentences to sound collaborative and solution-focused rather than apologetic.
> Use active verbs, remove corporate transition words like

'moreover' and 'therefore,' and keep the entire message under four sentences so it is easy to read on a mobile phone.

Why this is the best: This is a master-class follow-up. It uses the three-part feedback structure perfectly.

It praises the good part (the technical explanation), cuts the bad part (the defensive, apologetic tone), and guides the next attempt with specific rules (active verbs, no transition words, strict length limit of four sentences).

The resulting email will be sharp, professional, and perfectly tailored to get a positive response from a busy colleague.

Try It Now Exercise (5-10 Minutes)

Open your AI tool of choice, and let us practice the follow-up loop right now. This simple exercise will show you how quickly you can turn a weak first draft into a useful win.

1. **The First Prompt**: Ask the AI to write a short paragraph explaining a topic you know well (such as your favorite hobby, your professional specialty, or a basic household chore). Keep it simple: "Explain how to maintain a backyard garden."

2. **The First Output**: Read the response. It will likely be a decent but dry, generic list of tips.

3. **The First Follow-Up (The Jargon Cut)**: Type this exact message into the chat: > "This is too detailed and sounds like a textbook. Rewrite it in a warm, friendly tone. Use a simple, everyday analogy to explain the main point, and keep it under 100 words."

4. **The Second Follow-Up (The Concrete Example)**: Once the AI responds, push it one step further: > "Great. Now, add one specific, real-world example of a common mistake a beginner makes when doing this, and tell me how to avoid it in two sentences."

Compare the first response to your final version. You will see how much more engaging and useful the text became after just two follow-up questions.

Prompt Rescue Box

The Problem

The AI has completely lost track of your original goal. It got confused by your follow-up requests and went off on a bizarre, useless tangent, or it started repeating its own mistakes over and over.

The Rescue Prompt

Don't give up. Copy and paste this exact message into the chat to pull your assistant back to the main path:

> Let's pause and reset. We are going off track. Let's go back to my previous request. Please focus only on [state your specific, original goal here] and ignore the last change I asked for. Keep your response simple, brief, and directly focused on that one task." This reset works because it helps the AI ignore recent distractions and focus on your main goal.

Human-in-Charge Check

When you use follow-up questions to refine a draft, never assume that the AI's suggestions are correct.

Sometimes, in its effort to please you, the AI will make wild, unrealistic assumptions or invent completely untrue details.

Always stay in control. If a follow-up feels unnatural, untrue, or risky at work, use the Reset Prompt and guide the conversation in a safer direction.

Workbook Prompt

Take a moment to sit with a pen and a piece of paper, and answer this simple question honestly:

> When an AI tool gives me a weak or generic first response, what do I usually do? Do I give up and close the tab, or do I treat it as a draft and keep working to improve it?

Carol Higgins stood in her quiet, warm kitchen in Sarasota, her small carry-on suitcase resting neatly by the back door.

The morning sun was beginning to break through the palm trees, casting long, golden shadows across the linoleum floor. Barnaby sat near the door, watching her with sleepy eyes.

Carol adjusted her cotton t-shirt, draped her soft cardigan over her shoulders, and carried her medium jacket. She felt comfortable, light, and completely prepared for both the Florida heat and the cool Denver breeze waiting for her across the country.

She had not let her fear of the empty chat box stop her. She had not accepted a heavy, useless fifty-pound suitcase just because a computer program gave her a bad list on its first try.

She kept the conversation going. She guided her assistant, asked for changes, and shaped the answer until it fit her life. She was truly in charge of her own journey.

She picked up her bag, feeling its light, manageable weight. She smiled, stepped out into the warm morning air, and closed the door behind her.

But as Carol began her journey to the airport, her mind drifted to others who might be struggling with these new tools.

She had found a way to use follow-up questions to fix her packing list. It had been a friendly, cooperative discussion.

But what happens when the conversation doesn't go smoothly?

What happens when the AI doesn't just need a little nudge, but gives an answer that is completely, spectacularly broken?

What do we do when our digital assistant gets stuck in a loop of nonsense, makes up fake facts with absolute confidence, or refuses to listen to our instructions at all?

To handle those moments without losing our temper or throwing our computers out the window, we need more than just friendly questions. We need a systematic, step-by-step diagnostic toolkit.

We need a reliable way to spot exactly why a prompt went wrong, and a set of emergency scripts to rescue a completely crashed conversation on the fly.

Chapter Five

The Prompt Rescue System

What to Do When AI Gives a Bad Answer

The streets of London were filled with the sound of brass bands, shouting crowds, and the crackle of fireworks on the night of August 16, 1858.

Hundreds of gas lamps lit up every public building. People shook hands with strangers in the streets, and newspapers announced that a huge obstacle in human history had finally been overcome.

Across the Atlantic Ocean, in New York City, the celebration was even louder. A hundred guns fired in City Hall Park, bells rang from every church tower, and a massive parade stretched for miles down Broadway.

The cause of this joy was a thin, copper wire stretching nearly two thousand miles across the cold, dark floor of the Atlantic Ocean. For the first time, North America and Europe were connected by a telegraph cable.

To mark the occasion, Queen Victoria sent a ninety-nine-word message of peace to President James Buchanan. The message was a heartfelt statement about friendship between nations.

To the outside world, it looked like a major engineering victory. Suddenly, the world felt much smaller.

But inside a small, damp wooden station on the coast of Newfoundland, Canada, things were very different. There was a quiet sense of panic.

An operator sat in a dark room, his eyes strained as he stared at a small brass needle. The needle was supposed to swing back and forth to signal the letters of the Queen's message.

But the needle hardly moved. It shook and twitched, sending out weak, blurry signals that were nearly impossible to read.

The Queen's ninety-nine-word message didn't come through right away. It took sixteen and a half hours of slow, repetitive work to translate the weak, distorted signals moving through the wire. Using our structured seven-step system can help you diagnose and fix such issues efficiently, giving you confidence in resolving common AI response problems.

The man responsible for the cable's electrical system was Wildman Whitehouse, a doctor who became an engineer. He was very confident and believed that if a technology failed, you just needed to push it harder.

He didn't stop to ask why the signal was weak. He didn't check the copper wire's structure or consider how water might affect the electrical current.

Instead, he tried to force the signal through by increasing the power.

Whitehouse hooked up huge, five-foot-long induction coils to the cable. He sent up to 2,000 volts of electricity through the delicate copper wire, hoping to make the weak signals easier to read.

For a few days, the high voltage seemed to work. A few brief messages scrambled across the ocean.

But behind the scenes, the high voltage was quietly burning through the gutta-percha insulation that kept the copper wire safe from the cold seawater.

Within three weeks of the big opening, the signals became totally garbled. They turned into a mess of random twitches and empty gaps.

By October, the cable had stopped working completely. The copper wire was ruined by the same electricity that was supposed to fix it.

The great transatlantic cable didn't fail because the ocean was too wide or the telegraph was a bad idea. It failed because an engineer tried to fix a communication problem by pushing harder instead of addressing the root cause.

Sam Vance's Quiet Office Storm

A similar storm was brewing in the small back office of Iron & Oak in East Nashville, Tennessee.

It was 9:30 on a Thursday evening. From the outside, the gym looked perfectly peaceful.

The heavy iron weights were stacked neatly on their racks. The black rubber floor was swept clean. The warm string lights hanging from the old wooden rafters cast a soft, golden glow over the empty workout space.

But inside the small office, Sam Vance was quietly frustrated.

He sat at his wooden desk, his face illuminated by the cold blue light of his laptop screen. Next to his computer lay a cold, half-eaten slice of pizza and a crumpled paper cup.

Sam was working on an email campaign to invite local parents to his gym's autumn strength class. If he could get fifteen parents to sign up, he'd be able to pay off his equipment loans for the year and finally get a good night's sleep.

He tried using his AI assistant to draft the emails, hoping it would save him time. But the results were terrible because he didn't recognize common issues like vague prompts or misaligned tone, which can cause poor responses.

His first prompt had been simple: Write a promotional email for a gym.

The AI had responded with a long, aggressive sales pitch. It was packed with phrases like "Unleash your inner warrior!" and "No pain, no gain!"

Sam cringed. He didn't run a sweaty boot camp. His gym was a welcoming place for parents who wanted to play with their kids without hurting their knees.

He tried to fix the AI's response. But instead of figuring out what was wrong, he made the same mistake as Wildman Whitehouse with the Atlantic cable. Addressing the root cause of issues, rather than just pushing harder, can help you troubleshoot more effectively and feel more in control of your AI responses.

He typed: No, make it sound more exciting! Make it sell better! But without specific guidance, the AI produced only louder, more exaggerated responses, underscoring the need for clearer, more detailed prompts to achieve better results.

The AI changed its response, but it only got louder and more robotic. It started using lots of exclamation marks, capital letters, and even cheesier marketing cliches. Employing targeted emergency scripts for common problems, such as overly long or generic responses, can help you quickly regain control and improve your results.

> Are you tired of feeling sluggish? Embark on a life-changing fitness journey today! Don't miss out on this gold-standard opportunity to unlock your ultimate potential! Act now before it is too late!

"This is garbage," Sam muttered, hitting his hand against the desk.

He tried one more time, his fingers flying across the keyboard in frustration: Make it shorter and stop sounding like a commercial!!!

The AI then gave him a two-sentence draft that was so dry and cold, it sounded like a legal notice:

> We have a strength class starting on Saturday. You should sign up today to avoid missing your spot.

Sam closed his eyes and leaned back in his chair. He felt completely defeated.

He had spent two hours typing, correcting, and getting frustrated with his computer. The more he pushed the AI, the worse and less useful the results became.

He remembered the newsletter he needed to send to his five hundred local subscribers the next morning. He worried about his loans. It felt like he was dealing with a broken machine that promised to help but only wasted his time.

Sam was ready to delete his account and write the emails himself, even if it took him until three in the morning. He didn't realize the AI wasn't actually broken.

The problem was that his prompt lacked sufficient structure. He was trying to fix a poor communication channel by simply repeating his demands with greater frustration. What he really needed was to stop pushing and start fixing the prompt.

Why Bad AI Answers Happen

When an artificial intelligence tool gives you a bad, robotic, or useless answer, it is easy to assume that the technology has failed.

But as the person in charge, you need to look beyond what's on the screen. It's important to understand why the digital signal broke down in the first place.

AI does not generate bad answers because it dislikes you or because it wants to be difficult. It generates bad answers because of a few simple, predictable structural errors in your communication:

1. **The Vague Goal**: You asked the AI to "help" or "write something," without defining the specific physical action or the task's scope.

2. **The Missing Audience**: You did not tell the AI who would read the final text, so it defaulted to writing for the average, neutral internet user.

3. **The Cliché Default**: Because public models are trained on billions of pages of marketing text, they naturally default to using lazy, high-pressure clichés unless you explicitly tell them to stop.

4. **The Task Avalanche**: You tried to ask the AI to handle a massive, multi-step project (like writing an entire email campaign, a landing page, and a schedule) in a single, giant prompt.

5. **The Unstated Format**: You did not provide a visual structure, so the AI delivered a massive, dense wall of text that was too tiring to read.

6. **The Blind Guess**: The AI lacked sufficient background on your business, budget, and constraints, so it was forced to make a wild, average guess.

7. **The Screen Trap**: You accepted the very first draft without realizing that the first response is always a rough working copy that needs active, human guidance.

Once you know these seven reasons for bad answers, you don't have to feel frustrated or helpless anymore. You don't need to buy a huge database of prompts.

You need a simple system to figure out what went wrong and fix the problem as you go.

The Seven-Step Prompt Rescue System

The **Prompt Rescue System** is a reliable, step-by-step toolkit you can use to fix any conversation that's going wrong with an AI tool.

Instead of starting over, you can use these seven steps to spot the problem, fix your prompt, and get your digital assistant back on track:

Is the AI's response...

- Unclear or Unstructured? - Add a clear Goal and Format constraint.
- Generic or Bland? - Define your specific Audience and Context.
- Factually Suspicious? - Ask for verification steps and sources.
- Robotic or Cliché? - Request a warm tone and plain language.
- Too Complex or Broad? - Break the task down into smaller prompts.

Step 1: Name the Problem

Look at the AI's bad response and identify exactly what is wrong. Is it too long? Is it too formal? Does it sound like a pushy salesperson? Is it missing key facts?

Naming the exact problem keeps you from typing vague corrections like "make it better."

Step 2: Restate the Goal

Refocus the conversation on the exact result you want. Tell the AI exactly what you need it to create (for example, "We are writing a three-paragraph invitation email").

Step 3: Add Missing Context

Figure out what background details the AI doesn't know. Tell it who your audience is, what your business values are, or any limits you have.

Step 4: Narrow the Task

If the project is too big, break it into smaller parts. Don't ask for a whole campaign at once. Ask the AI to write just the first email or just some subject line options before moving on.

Step 5: Specify the Format

Ask for a clear visual structure. Request short, one-sentence paragraphs, a bulleted list, or a simple table to make the text easy to read.

Step 6: Set the Quality Standard

Tell the AI which words, clichés, or tones to avoid. Specifically ban words like delve, testament, unleash, and embark.

Step 7: Verify and Human-Edit

Take the improved draft, read it out loud, and make any final personal edits on your own computer.

Emergency Rescue Scripts by Problem

When you get a specific kind of bad AI answer, you don't have to come up with a fix from scratch. You can copy, paste, and adjust these targeted emergency scripts to fix the problem quickly:

When the Output Is Too Generic

This happens when the AI gives you broad, generic advice that doesn't help at all.

> This response is too general. Let us make it highly specific. Please rewrite this for [describe your specific audience/situation]. Add three concrete, real-world examples of how this works, and remove any advice that could apply to any general business.

When the Output Sounds Too Robotic

This happens when the AI uses very formal language, complicated words, or sounds like a machine pretending to have feelings.

> This draft sounds stiff and unnatural. Please rewrite it to sound like a warm, down-to-earth conversation between two neighbors. Use simple, everyday language, shorter sentences, and a supportive tone. Avoid any marketing hype, exclamation marks, or formal business transition words.

When the Output Is Too Long

This happens when the AI gives you a big block of text that's too dense to read on your phone.

> This response is far too long for what I need. Please cut this draft by 50%. Keep only the most important points, organize them into short, bulleted lists, and make sure the entire response can be read in under sixty seconds.

When the Output Is Too Short

This happens when the AI gives you a short, lazy answer that doesn't have enough detail or useful information.

> This is a good start, but it is too brief. Please expand on this draft. Add step-by-step instructions for how to complete the task, list two common mistakes to watch out for, and explain the main point in more detail.

When the Output Is Factually Suspicious

This happens when the AI gives you statistics, quotes, or historical facts that seem a bit off or possibly made up.

> I want to double-check the accuracy of these facts. Please list every historical claim, date, or statistic you used in your previous response. For each one, please tell me what specific, official website or primary source I should check to verify its accuracy. If you do not have access to live data, state that clearly.

When the Output Goes Off-Topic

This happens when the AI gets confused by earlier messages and starts talking about something totally unrelated to your goal.

> We are going off-track. Let us pause and return to our original goal: [state your original goal]. Please ignore my last few messages, strip away any extra ideas, and give me a clean, simple draft focused only on that one task.

When the Output Gives Too Many Options

This happens when you ask for advice, and the AI gives you 30 different suggestions, leaving you feeling overwhelmed.

> This is too many options to consider. Please analyze these suggestions and rank them from most practical to least practical for [your situation]. Give me only your top three choices, and explain the main benefit and risk of each one in two sentences.

The Before-and-After Rescue Process

To see how the Prompt Rescue System works, let's look at a real example of improving a gym landing page.

The Original Vague Prompt

> Write a sales page for a gym.

The Weak AI Output

> Welcome to the ultimate fitness experience! Are you ready to crush your goals and unleash the beast within? Our state-of-the-art facility is a testament to our core strength. Embark on a life-changing journey today! Leverage our synergy to maximize your potential!

The Diagnosis

- **The Problem**: It's full of aggressive marketing hype, corporate buzzwords (testament, embark, leverage, synergy), and sounds totally unnatural.
- **The Missing Pieces**: There's no specific audience, no context about the gym's values, no clear formatting, and no quality standards.

The Rescue Prompt

Sam uses the Prompt Rescue System to write a focused correction:

> Let us pause and rescue this draft. The tone sounds too aggressive and uses too many cheesy marketing cliches.
> Let's rewrite this sales copy for busy parents over forty who feel intimidated by loud, sweaty gyms. Focus on stress relief, gentle strength building to protect joints, and a welcoming, quiet community.
> Please deliver the draft in this format: use a short, three-sentence introduction, followed by a bulleted list of three specific benefits, and end with a friendly call to action. Do not use words like 'unleash,' 'beast,' 'embark,' 'leverage,' or 'synergy.'

The Improved Output

> **A Welcoming Space to Get Strong and Move Without Pain**
> We know that walking into a loud, crowded gym can feel overwhelming, especially if your joints already ache from a

elming, especially if your joints already ache from a busy week. At Iron & Oak, we believe that fitness should support your life, not exhaust you. Our quiet, supportive space is designed to help you build practical strength at your own pace.

- **Gentle Strength Building**: We focus on safe, functional movements that protect your knees, shoulders, and lower back.
- **A Patient, Expert Coach**: You will never be left to figure things out on your own; a coach is always beside you to guide your form.
- **A Quiet, Friendly Community**: There are no loud mirrors, screaming music, or high-pressure expectations here.

Want to see if we are a good fit for you? Join us this Saturday at 9:00 a.m. for a free, beginner-friendly class. We would love to meet you and share a cup of coffee afterward.

Sam's Breakthrough

Let's go back to the quiet office at Iron & Oak in East Nashville.

The old grandfather clock in the gym lobby chimed 10:15 p.m. Sam Vance sat in his chair, but his shoulders were finally relaxed. He had stopped hitting his hand on the desk.

After reading about the Prompt Rescue System, Sam saw where he went wrong. He had acted just like Wildman Whitehouse, pushing his digital assistant harder instead of fixing the real problem with his prompt.

He opened his chat window, took a deep breath, and looked over his earlier, frustrated prompts. He realized the problem: the AI was using loud, pushy sales language because he hadn't defined his audience or told it to avoid cliches.

He decided to run an emergency rescue prompt. He typed:

> Let's pause and reset this conversation. The emails you drafted sound too aggressive and sales-heavy. I want to start over with a fresh, clear approach.
> Our goal is to write a single, warm invitation email for our upcoming 'Saturday Morning Strength' class. The audience is busy local parents over forty who want to get strong but are intimidated by hardcore gyms.
> Please write this draft using a friendly, supportive, and neighborly tone. Use short, one-sentence paragraphs so it is easy to read on a phone.
> Do not use any exclamation marks, high-pressure sales tactics, or cliches like 'embark,' 'unleash,' 'no excuses,' or 'crush your goals.'

Sam hit send and watched the cursor blink.

Within seconds, the AI responded. The text flowed onto the screen, clean, quiet, and completely transformed:

> **Subject**: A quiet place to get strong (and a free class this Saturday)
> Hi Felix,
> If you have ever felt out of place in a loud, crowded gym with mirrors and screaming music, you are not alone.
> Many of our members at Iron & Oak felt the same way before they walked through our doors.
> We built our gym to be a quiet, welcoming space for busy parents over forty who want to build strength without hurting their joints.
> This Saturday at 9:00 a.m., we are hosting a small, free strength class designed specifically for beginners.
> Our coaches will guide you through every movement slowly and safely, and we will have fresh coffee waiting for you afterward.
> If you would like to join us, reply to this email and let me know. We would love to have you.
> Warmly,

Sam Vance
Owner, Iron & Oak

Sam read the email twice. He felt a sudden wave of relief.

"This is it," he whispered. "It sounds like a real person. It finally sounds like me."

He highlighted the text, copied it, and pasted it into his newsletter software.

Once the draft was in his newsletter system, he spent three minutes making a few personal edits. He added the gym's address, mentioned they'd be serving local coffee from the shop down the street, and scheduled the newsletter to go out at 8:00 a.m. the next morning.

He closed his laptop with a soft, satisfying click. Then he stood up, left his office, and turned off the gym's main lights.

As he locked the front door of Iron & Oak and stepped into the cool, quiet Nashville night, he felt truly empowered. He hadn't let frustration ruin his evening, and he hadn't settled for a bad, robotic draft that would have embarrassed him. He stayed in control. He had fixed the problem.

Try It Now Exercise (5-10 Minutes)

The best way to get good at the Prompt Rescue System is to practice fixing a purposely bad answer. Open your favorite AI tool and follow these simple steps:

1. **The Set-Up**: Intentionally input a vague, bad prompt that is guaranteed to produce a robotic, cliche-filled response: > "Write a sales pitch for a business."

2. **The Analysis**: Read the AI's terrible output. Count how many cliches (such as unleash, revolutionize, testament, or in today's world) appear in the text.

3. **The Rescue**: Copy and paste this exact rescue script, customizing the bracketed details: > "This response is far too sales-heavy and generic. Let us rescue this draft. Please rewrite this pitch for [describe your specific favorite business, hobby, or product]. Use a warm, conversational, and direct tone. Organize the information

into three short bullet points, and do not use any marketing cliches or buzzwords."

4. **The Compare**: See how quickly the AI drops the robotic sales pitch and gives you a clean, useful, and natural draft.

Prompt Rescue Decision Tree

Use this simple mental checklist to quickly figure out what's wrong with any weak AI response:

- **Is the response too long or packed with text?**
  Yes -> Run the Jargon Cutter script to cut the length by 50% and demand bullet points.

- **Is the response too dry, formal, or cold?**
  Yes -> Run the Robotic script to demand a warm, conversational neighborly tone.

- **Is the response too vague or academic?**
  Yes -> Run the Concrete Example script to ask for three real-world illustrations.

- **Is the response too large or confusing?**
  Yes -> Run the Off-Topic script to reset the goal and focus on one tiny task first.

Human-in-Charge Check

When you use the Prompt Rescue System, remember you're in charge. If an AI tool keeps giving you the wrong answer after three tries, don't keep typing in the same chat window.

The AI's memory might be cluttered with too many mixed instructions.

Take a step back, open a new chat window, and simplify your original prompt using the Five-Part Formula. You're the boss; you decide when it's time to start over.

Personal Prompt Rescue Log

Make a simple notebook page or digital document to keep track of your rescue wins and build your confidence:

- **My Original Prompt:**
  e.g., Write a gym email.

- **What Went Wrong?**
  Too loud, pushy cliches.

- **The Diagnostic Problem:**
  Missing target audience.

- **My Chosen Rescue Script:**
  Robotic/Generic Script.

- **The Real-World Outcome.**
  Warm, authentic invite.

Sam Vance drove home through the quiet streets of East Nashville, the dashboard lights of his truck reflecting softly in his eyes.

The anxiety that had bothered him all afternoon was gone. His email campaign was scheduled, his loans were under control, and he had learned how to turn a tech problem into a simple, repeatable win. He felt like he had finally tamed the wild beast of artificial intelligence.

But as he parked his truck and walked up his front steps, a quiet thought lingered in his mind.

The email draft he created was beautiful and authentic. He felt safe using it because he had removed any client details before pasting them in, protecting their privacy just as Paula had.

But during the rescue process, the AI listed three scientific studies claiming that lifting weights reduces chronic lower back pain in adults over 40. It gave those statistics with total confidence.

Sam wanted to use those statistics in his newsletter to show his clients that strength training was safe.

But how could he be sure those numbers were actually true?

What if the AI had just made up those statistics? What if the scientific studies mentioned didn't even exist?

He realized that getting a great, natural response was only half the job. If he wanted to stay in control as a professional, he needed to know when to trust the AI's claims, when to double-check the facts, and how to handle the tricky world of digital truth.

Chapter Six

Trust, Privacy, and Safety

How to Stay Human in Charge

In a quiet apartment in Hamburg, Germany, a man sat at a wooden table, looking at a single black book. He ran his fingers over the gold letters on its imitation-leather cover. The book smelled musty, and its pages were yellowed at the edges, filled with neat handwriting. To him, this book was more than old paper; it felt like a secret key to history.

The year was 1981, and the man was Gerd Heidemann, a reporter for a famous German magazine called Stern. He believed he was holding Adolf Hitler's private diary.

Over the next two years, that one notebook became a stack of sixty-two volumes. Heidemann's editors were excited. They thought they had found the greatest historical discovery of the century. The magazine paid more than 9 million German marks, the equivalent of millions of dollars today, to buy the books from a quiet collector who said they had been saved from a plane crash in 1945.

Before publishing the diaries, the editors knew they had to check them. They sent samples of the handwriting to famous historians.

One of these experts was Hugh Trevor-Roper, a respected professor of history at Oxford University. He traveled to a secure bank vault in Switzerland to examine the heavy volumes. He sat at a table, turning the pages, reading the personal thoughts of one of history's most infamous figures.

The handwriting looked right, and the signature matched the one on official documents. The diaries included small, ordinary details that seemed too real to be fake. Trevor-Roper got caught up in the excitement and said the diaries were genuine.

The news flashed across the globe. Newspapers in Britain and America paid fortunes to print the translated pages.

But then, a small group of scientists at the German Federal Archives decided to perform a different kind of check. They did not look at the handwriting or the historical facts. They did not read the sentences to see if they sounded real.

Instead, they took a tiny piece of the paper and put it under a special blue light.

Under the light, the paper glowed with a bright, blue-white radiance. This glow was caused by chemical "optical brighteners." These chemicals were first used to make paper look whiter in 1954, nine years after the end of World War II.

The scientists then tested the glue used in the bindings. It contained modern plastics that did not exist in the 1940s. Finally, they tested the ink. It was fresh, and it had been applied to the paper recently.

All sixty-two diaries were fake, written by an art forger in a basement. The gold letters on the cover, which people thought meant "Adolf Hitler," were actually "FH" in an old German font because the forger ran out of the letter "A." People were so eager to believe the story that they missed a simple mistake.

The Great Hitler Diaries Hoax worked because even smart people believed a confident story. They trusted the handwriting because it looked right, and they believed the story because it fit what they expected. They did not check the real evidence in front of them.

This historical mistake shows why many people have trouble using artificial intelligence safely today.

When you type a question into an AI tool, the screen fills with a beautifully written response. The sentences are grammatically flawless. The tone is calm and professional. The facts are presented with absolute, unwavering confidence.

This confidence is only an illusion. Like the gold letters on the fake diaries, the perfect look of an AI's answer can hide mistakes inside, reminding you to stay cautious and not take AI responses at face value.

To stay in control, you need to look past the machine's polished writing. Learn when to trust, when to double-check, and how to use simple tests, like verifying sources or cross-referencing facts, to protect yourself using today's versions of the optical brightener test.

Why AI Confidently Invents Things

To see why your digital assistant can sound correct but still be wrong, you need to know how it works.

An AI tool does not have a library of facts stored in its memory. It does not "know" things the way you do.

When you write a sentence, your brain pulls up memories, connects them to ideas, and tries to describe something real. If someone asks where you went to school, you picture your old classroom, remember the name, and answer honestly.

An AI does not have a classroom to picture. It is a language prediction model.

AI works by finding patterns in words. When you type a prompt, it reads your input and predicts which word should come next, using the billions of sentences it has seen before.

It's like a much smarter version of the auto-complete feature on your phone.

If you type "bread," your phone might suggest butter or water. It does not know what bread is or that you want breakfast. It suggests these words because they often appear together in English.

Because of this word-matching process, AI models suffer from three distinct truth-telling limitations:

1. The Confidence Trap

The AI does not have an "I don't know" state unless you explicitly train it to say so.

If you ask a human assistant a question they cannot answer, they will usually look confused, hesitate, and say they need to look it up.

But AI is designed to keep generating language. If it finds a gap in what it knows, it does not stop. It just keeps predicting the next word.

This leads to **hallucinations**, in which the AI confidently makes up names, dates, statistics, and events that sound real but are not true.

2. Fabricated Sources

This is maybe the most dangerous part of the confidence trap. If you ask an AI to show its sources, it does not search the live internet for real books or articles.

Instead, it will predict what a credible source citation should look like.

It will create a source that looks real, with an author name, book title, publisher, and even page numbers. But if you check in a real library or online, you will find the book does not exist. The AI made up the citation to meet your request, so always verify sources independently.

3. The Knowledge Gap

Most public AI models are trained on data up to a certain date. They do not know today's news or local changes in your area or business unless you tell them in your prompt.

If you ask for the best restaurant nearby, it might confidently suggest a place that closed two years ago.

Knowing these limits does not mean you should stop using AI. It just means you need to change your role. You are not just a passive reader anymore. You are the editor, the fact-checker, and the final judge of what is true.

Carol Higgins and the Dental Invoice Scare

To see how this verification process works in everyday life, let us return to Carol Higgins's sunny kitchen in Sarasota, Florida.

It was 10:00 a.m. on a Friday, and Carol was sitting at her table with a cup of chamomile tea. The morning light cast bright squares on the tiled floor, but Carol was not looking at the sun. She was staring at a single sheet of paper that had arrived in her mail.

It was an invoice from her dentist's office.

At the bottom of the page, printed in bold black ink, was the balance due: **$1,240.00**.

Carol's hand shook slightly as she set her teacup down. She had gone to the dentist three weeks ago for a routine crown replacement. Before the appointment, the receptionist had told her that her insurance would cover 80% of the cost. She had expected to pay a couple of hundred dollars, not a small fortune.

The invoice was filled with confusing abbreviations and dental codes: D2740 (Crown - Porcelain/Ceramic Substrate), D2950 (Core Buildup), and Deductible Non-Met.

Beneath the codes, a line read: Insurance adjustment: $0.00. Patient responsibility: 100%.

Carol felt a cold knot of panic in her stomach. She did not have an extra thousand dollars for one tooth. She wanted to call the dentist's office right away, but she knew she would feel lost when talking to the billing manager and would not know how to explain her side.

She picked up her tablet and opened her AI assistant. Her first instinct was to panic and let the machine take over. She wanted to type, "My dentist is scamming me." Help me write an angry letter to get out of paying this bill.

But she stopped herself. She remembered the **Human-in-Charge Principle**. She was in charge, and the AI was just her helper. She did not want her anger to prompt her to write a letter that could hurt her long-standing relationship with her dentist.

She also remembered that the AI did not know about her dental insurance policy and could not make legal decisions for her.

She took a deep breath, calmed her mind, and decided to use the Safe Summary Method. She typed a structured, conversational prompt into the box:

> Act as a patient, clear-thinking medical billing advocate. I need your help to understand an invoice from my dentist's office. I am a retired schoolteacher on a fixed income, and I want to understand these codes before I call the office to ask for a review.
>
> Here is our context: I was told my insurance would cover 80% of the cost of a crown replacement. However, the invoice shows that my insurance paid $0.00.

> Please explain the following codes in plain, simple English: D2740 and D2950.
> Tell me why an insurance company might refuse to pay for these specific procedures, and give me three polite, direct questions I can ask the dental receptionist to help resolve this bill calmly.

She pressed send. The AI responded within seconds, keeping its tone warm and supportive:

> "I understand how scary it is to receive an unexpected medical bill. Let's break this down together so you feel completely prepared for your call.
> **The Dental Codes Explained**:
> - **D2740 (Porcelain Crown)**: This is the actual cap that covers your damaged tooth. It is a very common procedure.
> - **D2950 (Core Buildup)**: This is where the dentist adds filling material to the tooth to make it strong enough to hold the new crown.
>
> **Why Insurance Paid $0.00 (Common Triggers)**:
> 1. **Prior Authorization**: Some insurance plans require the dentist to send X-rays before the procedure to prove the crown was medically necessary. If the office forgot to send these, the claim is automatically denied.
> 2. **Waiting Period**: If you recently changed insurance plans, there may be a six-month waiting period before they cover major dental work.
> 3. **The 'Missing Tooth' Rule**: If the crown is for a tooth that was missing before your policy started, they may not cover it.
>
> **Three Polite Questions to Ask**:
> 1. 'Could you please check if the insurance claim was denied because of a missing X-ray or a lack of prior authorization?'
> 2. 'Was this claim sent with the specific clinical notes showing that the tooth had a deep fracture and needed a core buildup (D2950) to support the crown?'
> 3. 'Can we submit an appeal to the insurance company with

> my digital X-rays to show them the medical necessity of the work?'"

Carol read the screen, her panic evaporating. The knot in her stomach untied.

The AI did not solve her problem or make the bill go away. But it acted as a translator. It took a confusing, scary document and turned it into a clear set of options.

She wrote the three questions down in her neat, schoolteacher handwriting on a pad of paper.

She picked up her phone, dialed the dentist's office, and asked to speak with the billing manager. When the manager answered, Carol did not sound angry or helpless. She read her first question slowly and calmly.

The billing manager paused on the other end of the line. "Oh," she said, her keyboard clicking rapidly. "Let me look at your file... I see what happened. We sent the claim for the crown, but our system forgot to attach the digital X-rays showing the fracture. The insurance company rejected it because they didn't have proof. I am so sorry about that."

The manager promised to resubmit the claim with the correct X-rays immediately. Two weeks later, Carol received a revised statement in the mail. Her insurance had paid its 80% share. Her final balance was just $248.00.

Carol solved her dental bill problem not because the AI had magic powers, but because she used it to learn. She checked the reasons behind the bill, stayed in control, and made the final call that fixed the mistake.

The Three-Level Trust Test

To protect yourself from confident mistakes and digital errors, you need a clear way to judge the information your AI assistant gives you. Not every response needs the same level of doubt.

If you spend three hours double-checking a simple recipe, you will waste your time. But if you spend zero minutes double-checking a legal document, you could ruin your career or your finances.

To stay safe and work efficiently, use the Three-Level Trust Test. This system puts your AI tasks into three risk levels and tells you how much checking each one needs:

THE THREE-LEVEL TRUST TEST

- High-Risk (Level 3) - Professional Verification Mandatory (Medical, Legal, Finance, Safety)

- Medium-Risk (Level 2) - Active Human Verification Required Summaries, Travel, Business Copy)

- Low-Risk (Level 1) - Verify Minimally / Edit for Style (Brainstorming, Drafts, Checklists)

Level 1: Low-Risk Uses (Verify minimally, edit for style)

These are tasks where a mistake might make you look a little silly, but it will not cause any real harm to your health, money, or feelings.

- **Examples**: Brainstorming creative angles for a blog post, drafting a basic invitation to a neighborhood potluck, organizing a messy list of grocery items, or writing a poem for a friend's birthday.

- **Action Step**: Read the output for tone and clarity, make sure it sounds natural, and use it freely. You do not need to fact-check if the poem uses the most accurate adjectives.

Level 2: Medium-Risk Uses (Active verification required)

These are tasks where a wrong fact or date could embarrass you at work, waste your time, or cause a small disagreement.

- **Examples**: Summarizing a long industry report for a team meeting, planning a travel itinerary with specific museum opening hours, writing product descriptions for your online shop, or comparing the features of two software tools.

- **Action Step**: Do not copy and paste the output mindlessly. Take five minutes to double-check key dates, product prices, and core factual claims against primary, trusted sources (such as official museum websites or product manual sheets) before sharing the work.

Level 3: High-Risk Uses (Professional verification mandatory)

These are tasks that affect your health, your money, your legal safety, or your company's future. A mistake here can have serious, life-changing results.

- **Examples**: Explaining the side effects of a new prescription drug, deciding how to invest your retirement savings, drafting a legal contract for a business partnership, or resolving a major insurance dispute.

- **Action Step**: Use the AI's answer only as a starting point to learn. Use it to understand the basics and to make a list of focused questions.

Always talk to a certified human expert, like a doctor, financial advisor, lawyer, or insurance agent, before making your final decision. Never let the machine make the final call about your safety.

What Never to Share

When you use your AI assistant, staying safe is not just about checking facts. It is also about protecting your own privacy.

Every time you paste text into a public AI tool, that data leaves your device and goes to servers you do not own or control.

To keep your personal, financial, and work life safe, remember these digital boundaries.

Never paste the following items into any public AI tool:

- **Corporate Passwords & PINs**: Never share any access keys, passwords, or security questions.

- **Official Government IDs**: Keep Social Security numbers, tax IDs, driver's licenses, and passport numbers out of the chat.

- **Sensitive Financial Data**: Never upload raw financial ledgers, credit card numbers, personal bank statements, or bank routing codes.
- **Private Patient Records**: Protect the medical histories, specific diagnoses, and names of family members, clients, or patients.
- **Proprietary Company Strategies**: Keep unreleased product plans, internal software code, and sensitive corporate files completely private.
- **Sensitive Legal Documents**: Do not paste active legal contracts, divorce papers, or dispute filings that contain real-world names and addresses.
- **Children's Information**: Never share names, school locations, or identifying details regarding children.

How to Fact-Check AI

To fact-check your digital assistant, you do not need to spend hours in a library. Just ask a few simple, focused questions about the AI's answer.

When you receive a response that contains historical claims, statistics, or scientific studies, use these three verification techniques:

1. Ask for the "Verification Map"

Do not ask the AI to prove its claim with a citation, as it might forge one.

Instead, ask it to tell you where a human should look. Type this prompt:

> What is the exact name of the official government agency, public library, or academic database where a human researcher can find the primary source for this statistic? Tell me what search terms I should type into that database to find the original document.

If the AI has invented the statistic, it will often confess in its next turn, stating that it lacks access to live data or that the number was an illustrative example.

2. Run the Cross-Model Check

If you are analyzing a complex historical or scientific topic, run your prompt through two different AI tools (such as Claude and ChatGPT).

Compare the two outputs side by side.

If both tools yield the same date, name, or statistic, the claim is more likely to be true. If they give different answers, that is a warning sign, and you should check the details using a regular search engine or an official website.

3. Check the Core Assumptions

Often, an AI gives a bad answer because it made a wrong guess about your question.

Force the assistant to show its work by typing:

> Please list the three most significant factual assumptions you made to generate this response. Tell me what details you assumed were true about my budget, my location, or my history.

AI and Modern Scams

As artificial intelligence becomes more powerful, some people use it to create very convincing scams.

Staying in control means protecting yourself not just in the chat box, but in all parts of your digital life.

Be aware of these four common AI-driven security risks:

1. Fake AI Mobile Apps

When you look for an AI tool in your phone's app store, you will see many apps with names that sound official. Many are copycats meant to trick you into paying high fees or stealing your personal data.

Always download AI tools directly from the official websites of established technology companies.

2. AI-Generated Phishing Emails

In the past, scam emails were easy to spot because they had spelling mistakes and bad grammar. Now, scammers use AI to write perfect, professional messages that look like they are from your bank, utility company, or employer.

They may even customize the email with details they found on your social media profiles.

Always verify the sender's actual email address, and never click on a link in an unexpected email without calling the company directly using a trusted phone number.

3. Voice Cloning Scams

This is a scary trick in which a scammer uses a short, three-second audio clip of a family member's voice, often from a public video, and feeds it into an AI tool.

The scammer then calls you, using a fake voice to pretend your relative is in trouble and needs money right away.

If you receive a suspicious call from a loved one asking for money, hang up immediately and call them back on their regular, trusted phone number.

Set up a family password, a simple secret word only your close family knows, to check their identity in an emergency.

4. Deepfake Photos and Videos

AI can now make very realistic photos and videos of real people doing or saying things they never actually did.

Be very careful with any surprising video or photo you see on social media that shows a public figure or family member saying shocking things, especially if it asks you to send money or click a link.

Try It Now Exercise (5–10 Minutes)

Let's practice fact-checking now. This simple exercise will show how easily an AI can make a mistake and how quickly you can catch it:

1. **The Prompt**: Open your AI tool and ask it to list three interesting historical facts about your hometown or the city where you currently live. Keep the prompt simple: "Tell me three interesting historical facts about Hannah."

2. **The Verification**: Read the response. Look at the names of the historical figures, the dates of events, and the names of old buildings.

3. **The Search**: Open a traditional search engine in a separate window. Search for one of the specific claims the AI made (such as "When was Finn built in Helen?").

4. **The Comparison**: Compare the official local history website with the AI's response. Did the AI get the exact year right? Did it blend two different historical figures into one?

See if there are any small mistakes or exaggerations in the AI's answer.

Prompt Rescue Box

The Problem

You think the AI has made up a statistic, a quote, or a book citation to support its point, but it did it so confidently that you are not sure.

The Rescue Prompt

Do not let the assistant get away with it. Copy and paste this message into the chat:

> Are these statistics, quotes, and sources completely real, or are they placeholders? If you do not have direct access to the live database or the primary document, state that clearly. Do not guess. If you are unsure about any figure, please tell me what credible, official agency website I should visit to find the actual, verified numbers myself.

This forces the AI to break character, admit its limitations, and point you to verified sources rather than hallucinating plausible-sounding lies.

Human-in-Charge Check

Always remember the main rule of modern technology: **The AI can analyze, draft, and summarize, but it cannot be responsible for the results.**

If you send a client an email with a fake statistic, you cannot blame the computer. If you sign a contract with a hidden, risky clause, the court will not care that an AI wrote it.

You are fully responsible for any action you take based on AI advice. You are in charge. Stay alert and pay attention.

Before I Trust the Answer

Use this checklist before acting on any information from an AI tool to protect your safety, privacy, and reputation:

☐ **Have I classified this task's risk level?** (Is this a Low, Medium, or High-risk task?
☐ **Did I make sure no private data was shared?** (Have I removed all passwords, SSNs, credit cards, real names, and company secrets from the prompts?)
☐ **Have I verified key dates, figures, and claims?** (Did I cross-reference critical details against a trusted primary source?)
☐ **Did I check for fabricated citations?** (Am I sure the books, studies, or sources listed by the AI actually exist in the real world?)
☐ **Does this response make common-sense human sense?** (Have I read the text out loud to make sure it doesn't sound suspicious or absurd?)
☐ **Have I consulted a certified professional for high-risk decisions?** (If this involves my health, finances, or legal safety, have I spoken to a human expert?)

By using the Three-Level Trust Test and learning to look past her assistant's confident writing, Carol Higgins turned a scary thousand-dollar bill into a small, stress-free fix. She protected her money, stayed in control, and showed she could handle the digital age with confidence.

But as Carol sat back and sipped her warm tea, she grew curious.

She had learned to use these tools safely at her kitchen table for her personal life. She knew how to protect her family's privacy and check her travel plans.

But what about the fast-paced, high-pressure world of business?

How could a professional like Paula Jenkins use these same safety and conversation tips to handle her busy Monday mornings at work? How could she use AI to write emails, summarize meetings, manage projects, and make important decisions under pressure, without losing her own voice or risking her company's private data?

To do that, she would need to move from the kitchen to the boardroom. She would have to turn these conversation skills into practical tools for the workplace to handle the challenges of office life.

Chapter Seven

AI for Work

Emails, Meetings, Projects, and Decisions

At 10:00 a.m. on a cool Monday morning, Paula Jenkins sat in her Raleigh office with a cup of hazelnut coffee. Her desk was tidy, and her email inbox was organized with no unread messages.

Her manager, David, stopped by her door with a stack of folders. He noticed her clean desk and relaxed posture, and shook his head in surprise.

"Paula, I don't know how you did it," David said, leaning on the doorframe. "Friday's emergency meeting left us with two hours of messy transcripts, three missed vendor deadlines, and a client who almost left. I thought you'd be buried in paperwork until noon. But you already sent the updated project plan, organized the action items, and wrote an email that calmed the client. How are you so far ahead?"

Paula smiled and took a sip of her coffee. "I had a very helpful assistant working with me this morning," she said softly.

Just two hours earlier, Paula's morning had seemed like a disaster.

When she got to her desk at 7:45 a.m., she felt a familiar knot of dread in her stomach. Friday's meeting had been full of scattered ideas, complaints, and unfinished plans. The transcript was over twenty pages long.

She also needed to write a sensitive email to their main client, explaining why the training program was delayed by two weeks. If she sounded defensive, the client might leave. If she sounded too soft, her company would seem weak.

To make matters worse, she had to create a new project schedule to make up for the lost time. In the past, this kind of work would have taken her all day. She would have skipped lunch, felt stressed, and gone home exhausted.

She almost did what her colleague Mark usually did.

The Stopwatch and the Chart

Mark was the Project Operations Manager for the company's Chicago branch. He was a smart, hard-working man who prided himself on his raw speed. He arrived early, stayed late, and treated every administrative task like a race.

When Mark opened his email inbox on Monday morning, he did not stop to think or plan. He wanted to get the work done as fast as possible. He opened an artificial intelligence tool, copied the entire, messy Friday meeting transcript, and pasted it into the chat box.

He typed a quick, rushed prompt: Summarize this meeting and write an update email to the team.

The AI did exactly what he asked. It generated a quick, dry summary and wrote a cold, formal email. Mark did not take the time to read the text or check its tone. He copied the output, pasted it into his email client, and blasted it out to his entire team of fifteen trainers.

Within thirty minutes, Mark's phone began to ring.

His lead trainer was upset because the email made it sound as if she were being blamed for the project delay. His junior coordinator was insulted by the message's cold, robotic tone. The client called his manager, complaining that the company sounded like a faceless corporation that did not care about their partnership.

Mark spent the rest of his Monday morning apologizing, making urgent phone calls, and trying to fix the problems. He used the technology to work faster, but it only created more work for him. He let the machine take over, and it led to trouble.

The difference between Paula's calm morning and Mark's chaos isn't new. It's a modern version of a struggle that started over a hundred years ago during the Industrial Revolution.

In the early 1900s, an engineer named Frederick Winslow Taylor introduced a system called "scientific management." Taylor believed that human beings in factories should be treated exactly like parts of a machine. He stood over workers with a stopwatch, measuring their movements down to the second. He wanted to strip away all personal choice, creativity, and human warmth in the name of pure speed.

Taylor's system made factories faster, but it had a big downside. It hurt workers' morale. People felt like robots, became unhappy, and lost trust. Factories then faced strikes, mistakes, and high turnover.

But another engineer of that era, Henry Gantt, saw a better path.

Gantt worked with Taylor, but he cared deeply about people. He saw that a workplace couldn't run on speed alone. He believed real efficiency came from teamwork, clear communication, and trust.

Gantt created a simple visual tool we still use today: the Gantt chart. He didn't make it to the time workers, but to help people see how their work connects. It was meant to support teamwork, not replace it.

Gantt showed that when people have clear structures and feel respected, they work well together. Real efficiency isn't about making people act like machines. It's about using good tools to help people communicate.

When Paula sat at her desk on Monday, she decided not to repeat Taylor's mistake. She knew that if she used her AI assistant to create quick, unedited drafts, she'd end up like Mark-stressed, isolated, and always apologizing for robotic errors.

She chose to follow Gantt's approach. She would use the AI to organize her thoughts, structure her plans, and clarify her choices, integrating it seamlessly into her workflow while ensuring her own voice guides the final decisions.

The Best Work Uses for AI

To manage a busy office without risking data breaches or compromising sensitive information, you must understand which tasks are safe to hand over to your digital assistant, and which ones require your active human judgment.

AI is great at handling routine tasks that often drain your energy, helping office workers feel more confident and empowered during their day.

When you let AI handle these routine tasks, you have more time to do what computers can't: build relationships, solve tricky problems, and show empathy to your team, while also reviewing AI outputs for professionalism and accuracy.

High-Value Office Tasks for AI

- **Polishing Email Tone**: You can write a quick, rough, or slightly frustrated draft of a message and ask the AI to rewrite it to sound polite, professional, and collaborative.
- **Simplifying Meeting Transcripts**: You can paste a long, messy transcript of a video call and ask the AI to extract a clean table of decisions made, action steps, owners, and deadlines.
- **Drafting Project Outlines**: If you are starting a new project, the AI can help you brainstorm the initial phases, common risks, and step-by-step checklists to keep you organized.
- **Structuring Decision Options**: When choosing between different vendors, software programs, or marketing strategies, the AI can build a clean comparison table to help you analyze your choices objectively.
- **Drafting Standard Operating Procedures (SOPs)**: You can explain a simple office process in plain language and ask the AI to turn it into a clear, numbered instruction guide for new employees.
- **Role-Playing Tough Conversations**: Before you walk into a difficult performance review or a salary negotiation, you can ask the AI to act as your boss or employee so that you can practice your talking points.

Email Prompts That Save Time

Writing daily emails takes up a lot of time at work. We spend hours typing, deleting, and rewriting messages, trying to sound both firm and friendly.

When you ask AI to help write an email, avoid giving it a vague prompt as Mark did.

If you type Write an email about a project delay, the AI will give you a generic, robotic form letter.

To get a clear and natural draft, give the AI four details: the message's purpose, your relationship with the reader, the tone you want, and the next step.

The Email Draft Prompt

Use this structured prompt template whenever you need to write a professional message:

> Act as a professional communication coach. I want to draft an email about [describe the situation safely].
> Here is our context: The recipient is [describe the reader, e.g., a valued client, a busy colleague, my manager]. My relationship with them is [e.g., very friendly, strictly professional, slightly tense]. The tone of this message must be [e.g., warm but firm, highly collaborative and supportive, direct and brief].
> Please deliver the draft using short, one-sentence paragraphs, so it is easy to read on a mobile phone. Give me three distinct options for the subject line at the very top, and include a clear, actionable next step at the end.
> Do not use any corporate jargon, defensive apologies, or clichés like 'delve' or 'please do not hesitate to contact me.'

By setting these boundaries, you stop the AI from sounding like a faceless company. You get a draft that feels personal and respectful, and that matches your goals.

Taming the Meeting Transcript

One of the hardest parts of office work is organizing the mess after a long meeting. We sit through long calls, take messy notes, and then struggle to remember who agreed to do what.

If your company uses software that records and transcribes meetings, you have access to a goldmine of data.

But a raw transcript is hard to read. It's full of interruptions, unfinished thoughts, and casual talk.

You can use your AI assistant to turn this messy text into a clear, useful summary of the project.

The Meeting Summary Prompt

To transform a messy transcript or a block of raw notes into a professional summary, copy and paste this prompt into your tool:

> Act as a meticulous administrative assistant. I am going to provide a raw transcript of a team meeting.
> Please read through the text and organize the information into a clean, professional summary. I want you to deliver the output in the following visual structure:
> **1. Executive Summary**: A brief, two-sentence explanation of the primary purpose of the meeting and the ultimate conclusion.
> **2. Key Decisions Made**: A bulleted list of the major agreements or choices made during the call.
> **3. The Action Steps Table**: A clean, four-column markdown table with these exact headers: **Task Description, Assigned Owner, Due Date,** and **First Step**.
> **4. Unresolved Issues**: A short list of questions or topics that were discussed but need further discussion.
> Here is the raw text: [paste your anonymized transcript or notes here]

This prompt helps the AI focus on what matters. It removes the casual talk and gives you a clear, organized summary you can quickly share with your team.

Project Planning Without the Dread

When you get a big new project at work, the hardest part is often knowing where to start. The project can feel overwhelming, so it's easy to put off taking action.

You can use your AI assistant to break the project into small, manageable tasks.

The AI can't do the work for you, but it can create a clear plan that makes organizing easier.

The Project Manager Prompt

Use this prompt to turn a large, intimidating goal into a step-by-step project schedule:

> Act as an expert project manager who specializes in clear, non-overwhelming workflows. I want to build a practical project plan for [describe the project, e.g., rolling out a new employee training program].
> Here is our context: We have a timeline of [e.g., four weeks] to complete this. The team consists of [e.g., myself and two trainers]. We want to avoid feeling rushed and identify the most common risks before we begin.
> Please deliver the plan in a clean, phase-by-phase format. For each phase, list:
> - The primary milestone or goal.
> - A step-by-step checklist of micro-tasks.
> - The single most critical risk for this phase and how to prevent it.
> - The very first action step I should take tomorrow morning is to build momentum.
> Before you finalize, list any assumptions you made about our resources or timeline."*

This plan is a visual guide. It doesn't replace your judgment; it just organizes the steps so you can focus on getting things done, one day at a time.

Making Decisions with Objectivity

As a professional, your value isn't about how fast you type. It's about the quality of your decisions.

But when we're stressed or tired, it's hard to weigh options fairly. We might pick the easiest path or get stuck overthinking because there are too many details to consider.

You can use your AI assistant to build a simple comparison tool that clearly shows the pros and cons of each option.

The Decision Matrix Prompt

When you are facing a difficult choice between different software vendors, strategies, or hiring candidates, run this prompt:

> Act as an objective, analytical business consultant. I am trying to decide about [describe the choice, e.g., which email marketing platform to buy for our gym].
> Here is our context: We are choosing between three options: [Option A, Option B, Option C]. Our primary constraints are [e.g., a $100 monthly budget, ease of use for beginners, and the ability to send automated welcome messages].
> Please help me compare these options. Create a clean Markdown table with the following columns: Option Name, Primary Benefit, Key Risk, Monthly Cost, Learning Curve (Low/Medium/High), and Best-Use Case.
> Directly below the table, write a brief, two-sentence recommendation for which option fits our specific constraints best, and explain your logical reasoning in simple language.

This comparison table removes the stress from decision-making. It shows the facts side by side, so you can confidently and clearly share your recommendation with your manager or team.

Workplace Boundaries and Safety

AI is a powerful tool for office work, but using it safely requires strict adherence to rules.

At work, you need to protect your own privacy and the reputation, legal safety, and trust of your company and clients.

1. The Corporate Data Lock

Never copy and paste proprietary code, unreleased product plans, internal financial ledgers, or private customer records into a public AI tool.

Many technology companies use publicly available prompts to train their future models. If you paste your company's secret code or client data, that information could eventually leak or be displayed to a competitor who types a similar query.

Always use the Safe Summary Method to strip away all specific corporate brands, client names, and financial figures before you seek help.

2. The Human-Touch Rule for HR

Never use artificial intelligence to make or deliver sensitive human resources decisions.

You should never use AI to write automated termination letters, disciplinary warnings, or performance rejections without heavy, active human editing and oversight.

Using a cold, computer-generated template for sensitive moments is disrespectful and risky, and it can cause legal problems. AI can handle codes, but only a person can show real compassion to an employee who is struggling.

3. The Verification Mandate

Never send a project update, a client report, or a financial projection generated by an AI without verifying every single date, name, and figure yourself on your own computer.

If the AI makes up a statistic or gives a wrong deadline in a plan you share, you are responsible for the mistake. The machine can't be fired, but you can.

Try It Now Exercise (5–10 Minutes)

Let's practice using AI to organize a messy block of office notes. This simple exercise will show how quickly you can turn chaos into a clear, professional summary:

1. **The Set-Up**: Imagine you just walked out of a messy team meeting. You have a quick, disjointed block of raw notes typed on your phone: > "Meeting about the new website rollout. John says the graphics aren't ready yet and need to be ready by next Wednesday. Sarah says the copywriting is done, but she wants to review it with the marketing team. We need to launch on the 15th of next month. Also, we forgot to discuss the ad budget.ds. Need to schedule a follow-up next Tuesday at 10 a.m."

2. **The Prompt**: Open your preferred AI tool, and paste this conversational instruction: > "Act as an organized project assistant. Please take my messy meeting notes and organize them into a clean, professional summary. > > Deliver the output in this format: use a short bulleted list of key decisions, followed by a simple table showing the **Task**, the **Owner**, and the **Deadline**. > > Here are the raw notes: [paste the messy notes from above]."

3. **The Review**: Read the output. See how the AI quickly cleaned up the notes and created a summary you can share with your team.

Prompt Rescue Box

The Problem

The email draft or project update from the AI sounds stiff and cold, or it uses too many corporate buzzwords like synergy, deliverable, or touch base.

The Rescue Prompt

Do not start a new chat. Paste this exact message into your existing conversation:

> This draft is too stiff and sounds like a plastic robot wrote it. Let's make it sound like a warm, professional, and collaborative human being.
> Please rewrite this message using simpler, direct English and shorter sentences. Remove any corporate jargon, buzzwords, and dramatic transitions.
> Keep the main facts, but ensure the tone is neighborly, respectful, and direct. Keep the entire email under five sentences.

Human-in-Charge Check

An email draft or project schedule is only as good as your final, human edit.

Before you copy any text from the AI and send it out, take a moment to read it out loud.

Ask yourself: Would I ever say these words in a real conversation with my colleague or client?

If a sentence sounds off or too formal, delete it and rewrite it in your own words. The AI is your assistant, but you are in charge.

Work Output Review

Use this quick checklist to review any document, email, or plan you create with AI before sharing it with your team or clients:

☐ **Have I verified all dates and deadlines?** (Are the dates listed in this plan accurate and realistic for our real-world schedule?)
☐ **Are all the facts and figures correct?** (Have I cross-referenced any statistics, metrics, or claims against my company's internal databases?)
☐ **Did I remove all private data?** (Have I made sure no proprietary software code, customer passwords, or financial ledgers were left in the text?)
☐ **Does the tone sound like my authentic voice?** (Have I removed any stiff corporate jargon, robotic clichés, or overly formal language?)
☐ **Is there a clear, actionable next step?** (Does the reader know exactly what they are supposed to do next after reading this message?)
☐ **Have I taken final responsibility for this work?** (Am I ready to stand behind this output as my own professional judgment, regardless of the tools I used to draft it?)

Paula's Monday Morning Triumph

Let's go back to the quiet office in Raleigh, North Carolina.

The clock showed 10:15 a.m. Paula Jenkins leaned back in her chair with a satisfied smile. Her hazelnut coffee was still warm.

Her morning meeting with David and the management team had gone very well.

When it was her turn to speak, Paula looked calm and confident. She opened her laptop and showed a clear table on the conference room screen.

The table showed what went wrong with the vendor, who was responsible for each next step, and how the team would recover the lost two weeks without extra hours.

Next, she shared the email draft she wrote to the client. The message had no corporate excuses or awkward apologies.

It was warm, honest, and collaborative. It explained the delay clearly, offered a solution, and invited the client to a short call to review the new schedule.

David looked at the screen, then at Paula with admiration. "This is incredible, Paula," he said. "The client already replied. They appreciate our honesty and clear plan, and they set up the call for this afternoon. You turned a crisis into a great example of professionalism."

Paula had thanked him, feeling confident.

She didn't let the Monday morning inbox overwhelm her. She didn't act like Mark, sending out cold, unedited drafts that caused confusion and frustration.

She stayed in control of her digital assistant. She used the Five-Part Prompt Formula to turn twenty pages of messy transcripts into a clear visual summary.

She stayed in charge, protected her company's data, saved hours of stressful work, and built her professional reputation. She was clearly the one in control of her career.

As she walked back to her desk, she felt calm. The digital future no longer seemed scary or overwhelming. Now, it felt like a clean, organized workspace where she could work with more ease, joy, and focus.

She sat down, ready to start her next task.

Her project schedule was set, her emails were organized, and her team was on the same page. But when she opened a blank document to write the introduction for the new employee training manual, she paused.

Organizing data and scheduling tasks were straightforward and logical. The AI was a great assistant for those jobs.

But writing was different.

To write the manual's introduction, she needed to do more than list facts. She wanted to inspire new employees and share her company's values, her experience, and her unique perspective.

When she asked the AI to write a blog post or an introduction before, it always produced a flat, dry draft full of clichés that sounded robotic.

How could she use her digital assistant to help write longer, more creative drafts without losing her own voice? How could she teach the AI to match her natural style, favorite words, and way of speaking, so her writing always sounded like her, not a computer?

To do that, she would need to move beyond the clear structures of project management and explore the world of creative writing. She would have to adjust her assistant's voice, avoid clichés, and learn how to draft together.

Chapter Eight

AI for Writing

Better Drafts Without Sounding Like a Robot

In the winter of 1852, a seventy-three-year-old retired doctor sat in a quiet study in London. His name was Peter Mark Roget. His hair was thin and white, and his eyes often grew tired after long hours of reading by candlelight.

Peter spent his life working in science and medicine. He studied how the lungs work, helped build a safe water system for London, and wrote thoughtful articles about natural science. But he struggled with sadness. His father died when he was young, and later, his wife passed away, leaving him to raise his children alone. To keep his mind from sinking into dark thoughts, Peter developed a quiet habit.

He began to make lists of words.

Describe how Peter's method of grouping words by ideas instead of alphabetically helps readers understand the importance of organizing thoughts to improve focus and engagement.

He didn't do this to become famous or to change literature. He did it to protect himself from his own thoughts. It was his personal coping tool. He believed that finding the right word for a thought could help him control his feelings. He wanted his writing and thinking to be clear.

For fifty years, Peter kept his lists in small leather notebooks. They sat on his shelves and gathered dust. He was very reluctant to share them with anyone. He worried people might think he was foolish or obsessive. After all, he was a scientist, not a poet.

But as he grew older, his friends noticed how clearly he spoke and how beautifully he wrote his letters. They urged him to publish his lists so that others could find the right words, too.

In 1852, he finally gave in and published his work. He called it Roget's Thesaurus. He did not expect much. He thought a few schoolteachers and writers might find it mildly useful.

But the book changed writing forever. Writers, poets, politicians, and everyday people rushed to buy it. It became one of the most successful books ever printed. It helped people move beyond repetitive language and find the exact words they wanted to use.

Peter Mark Roget never meant to change how people communicate. He was a reluctant pioneer who just wanted to clear his own mind. He knew that using the same words again and again makes our thoughts dull and weakens our human connection.

Explain how AI's reliance on the most common words causes blandness, emphasizing the importance of adding a human voice to keep content authentic and engaging.

To get past this fog, you need to use AI as a writing partner, not just as a shortcut, so your audience feels encouraged and confident in your voice.

Why AI Writing Sounds Like a Plastic Robot

To fix robotic writing, you first need to understand why it happens.

AI models are trained on billions of pages of publicly available text written by millions of people. When you ask an AI to write an article, an email, or a story, it does not sit down and think about its own life experiences. It does not have feelings, memories, or opinions.

Instead, it calculates which words are most likely to come next based on its training data. This process causes three main writing problems:

1. The Average Word Trap

Because AI guesses the next word based on what most people write, it always chooses the safest and most common phrases.

It is like a group of fifty people voting on what to eat for dinner. They will never agree on a spicy, unique dish. They will always choose plain cheese pizza because it is the one thing everyone can tolerate.

That's why AI writing often feels bland. It misses the creative and surprising choices that make human writing interesting.

2. Cliché Overuse

Public AI models often use predictable, repetitive phrases.

Because they have analyzed millions of marketing emails and business articles, they believe that professional writing must be packed with transition words and dramatic statements.

These overused words are like cardboard in digital form. Avoiding clichés helps your audience feel inspired to craft fresh, original writing.

These words are like cardboard in digital form. They fill space but add nothing special.

3. The Experience Gap

AI has never tasted a ripe peach on a hot summer day. It has never felt the quiet comfort of a sleeping pet on its lap, nor the sharp sting of losing a close friend.

Real human experiences and details make writing relatable, engaging, and honest, helping your audience feel connected and understood.

When AI writes about these topics without your help, it tends to use movie clichés and greeting-card phrases. It sounds fake because it isn't real.

Sam Vance's Kitchen Table Struggle

In his home in East Nashville, Tennessee, Sam Vance sat at his kitchen table on a quiet Tuesday evening.

His two children were fast asleep upstairs, and the only sound in the house was the gentle ticking of the wall clock. On his laptop screen, an empty document stared back at him.

Sam wanted to write a deep, encouraging blog post for his gym's weekly newsletter.

He wanted to share the story of a client named Maria. Maria was a forty-five-year-old mother of three who had walked into Iron & Oak six months ago, struggling with chronic knee pain. She had been terrified of lifting weights, believing she would hurt herself.

But with gentle guidance, she had built her strength, lost her pain, and recently sent Sam a photo of herself carrying her youngest child up a steep hill without any discomfort.

It was a touching story of victory. Sam wanted his newsletter to feel just as warm, but as he looked at the blank screen, his fingers felt heavy. He was tired from a long day at the gym, and his mind was empty.

He decided to ask his AI assistant to help him draft the story. He typed:

> Write a newsletter post about my client Maria. She had bad knees but started lifting weights at my gym, and now her pain is gone.

The AI responded instantly, delivering a draft that made Sam sigh in frustration:

> In today's fast-paced world, many individuals struggle with chronic joint discomfort. Maria embarked on a life-changing journey when she walked through the doors of Iron & Oak. Through the testament of our core methodology, she unlocked her ultimate potential. Moreover, lifting weights enabled her to adopt a pain-free lifestyle. Don't let excuses hold you back from your fitness journey!

Sam rubbed his temples. "This is terrible," he muttered. "Maria didn't 'embark on a journey' or 'unlock her potential.' She just wanted to pick up her toddler without her knees popping. And I would never use words like 'moreover' or 'delve' in my life. It sounds like a computer program trying to sell a miracle cure."

He thought about deleting the draft and spending two hours writing the newsletter himself, even though he was tired. He felt like the technology made him choose between speed and being real.

But then Sam remembered the Five-Part Prompt Formula. He realized he hadn't given the AI any style guidelines. He hadn't told it what his voice sounded like or banned the clichés.

He decided to use his writing toolkit and teach the AI to write more like a person.

The Writing Toolkit

To get better drafts from your AI assistant, stop treating it like a ghostwriter. Treat it as a writing partner instead.

Use these four practical tools to adjust the AI's voice, clean up its language, and keep your own perspective at the center.

Tool 1: The Voice Calibration System

The best way to make AI sound like you is to show it your own writing.

Instead of describing your voice with words like "friendly" or "professional," give the AI samples of your past writing and ask it to analyze your natural style.

The Voice Calibration Prompt

Open a fresh chat window and paste this exact prompt, attaching two or three paragraphs of your past writing (such as a well-received email, a blog post, or a personal letter):

> Act as a professional writing coach and style analyst.
> I want to train you on my personal writing voice so that you can help me draft future messages. Below, I have pasted two samples of my past writing.
> Please analyze these samples and describe my writing style in a simple bulleted list. Focus on:

1. My sentence structure (Do I use short, punchy sentences, or long, detailed ones?)

2. My vocabulary level (Do I use simple, everyday words, or formal, academic language?)

3. My emotional tone (Am I warm, direct, serious, humorous, or supportive?)

4. My transition style (How do I move from one idea to the next?)

5. Do not write any new drafts yet. Provide the analysis of my voice.

Here are my writing samples:
[Paste your sample 1 here]
[Paste your sample 2 here]

The Output and How to Use It

After the AI reviews your samples, it will give you a clear description of your unique voice.

It might say:

> Your style is direct and friendly. You use short, active sentences and simple, everyday words. You avoid business jargon and often start sentences with verbs to keep things moving.

You can use this description as your style guide. Whenever you ask the AI to write a draft, copy and paste this analysis into your prompt.

Tool 2: The Cliché Audit and Ban List

Robotic writing is easy to spot because it uses the same repetitive words.

You can quickly improve any AI draft by making a strict ban list of corporate buzzwords, transition words, and marketing clichés.

When writing prompts for emails, articles, or social media, copy and paste this specific quality check to the end of your instructions:

Apply this strict Cliché Ban List to your draft. Do not use any of the following words or phrases:

- Delve

- Testament

- Tapestry
- Moreover
- Therefore
- in conclusion
- in today's fast-paced world
- unleash your potential
- embark on a journey
- gold standard
- robust framework
- Please do not hesitate to contact me

Keep your language direct, conversational, and simple. Use verbs that show real, physical actions.

By removing these filler words, you make the AI write more clearly and directly.

The text will sound more human because it will use the simple, active verbs we use in real conversations.

Tool 3: Lived Experience Injection

AI can write with perfect grammar, but it lacks the real-world details that make writing feel true.

To make a story powerful and engaging, you need to give the AI the specific, messy, and emotional human details.

Think of this as the **Experience Seed**.

Before you ask the AI to write, jot down a quick list of your real memories, thoughts, and feelings about the topic. Don't worry about spelling or grammar; get your honest ideas on the page.

Example of an Experience Seed:

- **What happened**: Maria's knee pain was so bad that she couldn't pick up her toddler, Leo.
- **How she felt**: Scared, frustrated, felt like she was failing as a mom.
- **The turning point**: She lifted a 20-pound kettlebell for the first time. She was shaking, but she did it safely.
- **The result**: She sent me a photo of her carrying Leo up a hill in East Nashville, smiling. No pain.

How to Prompt with the Seed:

> Act as a warm, supportive writer. I want you to write a newsletter post based on a real-world experience.
> Here is the raw human context of what happened: [Paste your Experience Seed here].
> Please write a three-paragraph draft. Use my style analysis [insert your style analysis from Tool 1].
> Focus on the real, physical emotions of the situation. Start the post directly with the story, avoid any long-winded introductions, and do not use any clichés from our ban list.

This method ensures your final writing is grounded in real human truth.

The AI organizes the sentences, but the heart, emotions, and real details of the story are yours.

Tool 4: AI as the Editor, Not the Ghostwriter

Sometimes, the best way to use AI is to switch roles completely.

Instead of letting the AI write the first draft, write your own rough draft in your natural voice.

It doesn't matter if your draft is messy, disjointed, or has spelling mistakes.

Once your thoughts are on the page, give the draft to the AI and ask it to be an objective editor. This keeps your real voice at the center, while the AI helps polish the writing.

The Editor Prompt

> Act as a meticulous, encouraging copyeditor.
> I have written a rough draft of an article. Please help me improve its clarity, flow, and structure without changing my core meaning or losing my personal voice.
> Please perform the following tasks:
>
> 1. Identify any long, confusing sentences and suggest how to split them into shorter, punchier ones.
> 2. Highlight any repetitive words and suggest simpler alternatives.
> 3. Check the logical flow between paragraphs to ensure the reader can follow my thoughts easily.
> 4. Deliver your suggestions as a clean, bulleted list of specific changes, followed by a polished version of the text for me to review.
>
> Here is my rough draft:
> [Paste your rough draft here]

This way, you keep full ownership of your ideas.

You stay the main creator, and the AI acts as a quiet, helpful assistant who tidies up your work before you share it.

Before-and-After Writing Examples

Let's see how these four tools can turn a standard business draft into a warm, human message.

The Robotic Draft (Without the Toolkit)

> In today's fast-paced corporate landscape, it is a testament to our values to announce the upcoming volunteer day. We will delve into community service to leverage our core competencies in support of the local shelter. Moreover, embarking on this journey of service will unlock our collective potential. In conclusion, please do not hesitate to contact me to sign up!

Why this fails: It is packed with cold business clichés (today's fast-paced corporate landscape, testament, delve, leverage, moreover, embark, in conclusion). It sounds like a faceless corporation that is trying to check a box rather than a group of real people who want to help their neighbors.

The Human-Centered Draft (Using the Toolkit)

> Next Thursday afternoon, our office is closing early so we can spend a few hours helping out at the local family shelter.
> We are going to paint the community room, organize the food pantry shelves, and clean up the outdoor play area for the kids.
> This is a simple, practical way for us to support our neighbors, and we would love for everyone to join us.
> If you can help, reply to this email by Tuesday afternoon so we can get a headcount for carpooling. We'll have cold drinks and snacks waiting for everyone at the site.

Why this works: The language is direct and simple, and it describes real actions. It focuses on what people will actually do, not on abstract business ideas.

The tone is warm and friendly, with no corporate fluff. It sounds like a real person writing to coworkers.

Sam Vance's Newsletter Victory

Let's go back to the kitchen table in East Nashville.

It was now 10:30 p.m. The house was still quiet, but Sam Vance looked relaxed as he typed quickly on his keyboard.

After reading about the writing toolkit, Sam opened a new chat window. He realized his first try failed because he hadn't set any boundaries for the AI. He let the machine guess his voice.

First, he used Tool 1. He pasted two emails he had sent to clients, short, encouraging, and a bit blunt. The AI analyzed his style and noticed he used short paragraphs, active verbs, and a friendly, trainer-to-athlete tone.

Next, he used Tool 3. He wrote a quick, messy list of his real memories of Maria's journey: her fear of the weight room, her shaking hands during her first lift, and the photo she sent him from the park.

Finally, he combined these steps with the Cliché Ban List and hit send.

Within seconds, the AI delivered a brand-new draft:

> **Subject**: Maria's knees (and a photo that made my week)
>
> Six months ago, Maria walked into Iron & Oak with a look of pure fear on her face.
>
> Her knees had been hurting for over a year, and her doctor had told her to avoid lifting anything heavy. She was exhausted, frustrated, and sad because she couldn't pick up her toddler, Leo, without feeling a sharp pain.
>
> On her first day in the weight room, we didn't do anything crazy. We didn't lift heavy tires or scream through an intense workout.
>
> Instead, we focused on a simple, 20-pound kettlebell squat. Maria's hands were shaking, but she kept her back straight and completed the movement safely.
>
> Yesterday morning, I received a text message from her. It was a

photo of Maria carrying Leo on her shoulders as they walked up a steep trail at Shelby Park.
She was smiling, her knees felt great, and she wrote: 'Sam, I got my life back.'
Building strength isn't about looking like a fitness model. It is about having the physical freedom to enjoy your real life with the people you love.
Sam read the text on his screen, a slow, proud smile spreading across his face.

"This is absolutely perfect," he whispered. "It sounds like a real story. It sounds like something I would write."

He copied the text into his email newsletter software and spent two minutes making a few final edits. He added a sentence about the local weather, mentioned a group run at Shelby Park next Saturday, and clicked schedule.

He closed his laptop, feeling deeply satisfied.

He didn't let his exhaustion make him send a cheesy, robotic message. He didn't waste hours fighting writer's block. He used his digital assistant to organize his thoughts, but kept his own warm, human voice in charge of the story.

Try It Now Exercise (5–10 Minutes)

Let's practice training the AI on your unique voice right now. This simple exercise will show how quickly you can turn a robotic draft into something personal.

1. **Write a Rough Paragraph**: Spend three minutes writing a short, rough paragraph about something you care about (like your favorite childhood food, why you love your pet, or a simple tip for staying organized). Don't worry about style or grammar; use your own natural words.

2. **Analyze Your Style**: Open your AI tool and paste your paragraph with this instruction: > "Analyze this paragraph of my writing. Describe my tone, my sentence length, and my vocabulary style in three simple bullet points. Tell me what makes this text sound like

a real human."

3. **Draft a New Message**: Once the AI gives you the analysis, ask it to write a short, three-sentence email inviting a friend to dinner, using that exact style.
4. **Compare the Voice**: Read the invite the AI creates. Notice how much more natural and personal it sounds compared to a standard template.

Prompt Rescue Box

The Problem

The AI-generated draft is packed with annoying clichés, long-winded introductions, and transition words like 'delve,' 'testament,' 'moreover,' and 'in conclusion.'

The Rescue Prompt

Do not start a new chat. Copy and paste this exact message into your existing conversation:

> This draft is too robotic and uses too many filler words. Let's clean it up.
> Please rewrite this text. Remove all marketing clichés and transition words like 'delve,' 'testament,' 'moreover,' 'tapestry,' and 'in conclusion.'
> Use simple, direct, conversational language and shorter sentences. Start the message directly with the main point, and keep the entire text under 150 words."*

Human-in-Charge Check

The best way to test your writing is to read it out loud.

Before you publish, send, or print anything you wrote with AI, sit in a quiet room and read it out loud.

Listen to how the words sound together.

- If you find yourself tripping over a long sentence, split it in two.
- If a word sounds too formal or fancy for your normal vocabulary, swap it for a simpler word.
- If a paragraph makes you cringe or sounds like a corporate advertisement, delete it and write the main point in your own natural voice.

The AI is your assistant, but you are in charge of your own reputation.

Workbook Prompt

Take a moment to sit with a pen and a piece of paper, and answer this question honestly:

> When I write letters, emails, or stories in my daily life, what do I want my voice to sound like to my readers? Do I want to sound warm, direct, professional, funny, calm, simple, bold, or thoughtful? How can I protect those specific qualities when using digital tools?

Peter Mark Roget's small, handwritten notebooks became a bridge that helped millions of people find the right words to share.

He showed that writing isn't just repeating average phrases. It's a personal act of finding clarity and truth, and of sharing your unique view with the world.

By using the writing toolkit and guiding his assistant's voice, Sam Vance turned his weekly newsletter from a stressful task into a celebration of his community's wins.

He kept his unique voice, saved time, and proved we don't have to lose our humanity in a digital world.

But as Sam closed his laptop and walked to the stairs, he started thinking about a new challenge.

Writing a warm, encouraging story about his gym was creative and emotional. He had found a way to make the AI sound like a caring coach.

But what about times when we need to do the opposite?

What about when we face a huge, confusing wall of information we don't understand?

How can we use our digital assistant to help us learn, not just write?

How can we use AI to break down dense scientific articles, understand legal paperwork, study a new language, or research tough topics, without falling for fake facts or losing our critical thinking?

To do that, we need to move beyond creative writing and explore research and study. We must learn to use our digital assistant as a patient, plain-English tutor who helps us understand tough topics.

Chapter Nine

AI for Learning, Research, and Understanding Hard Things

The kitchen carried the scent of burnt toast mixed with the freshness of wet grass from the morning rain. The window felt cold, dotted with small drops of water. In the quiet room, the clock's tick-tock sounded especially loud.

Carol Higgins sat at her wooden table, tapping her fingers on a thick manila envelope. When she pulled out the paper inside, it made a dry, scratchy sound.

The letter was from her dental insurance company. It was a formal denial notice.

A few weeks ago, Carol had undergone a difficult dental procedure to save a broken tooth. Now, the insurance company was refusing to pay.

The letter was two pages, filled with small, dense black print. As Carol read, her throat went dry, and her heart beat faster.

The letter was packed with confusing terms like alternate benefit provisions, radiographical necessity, clinically unsubstantiated core build-up, and exclusionary period clauses. At the bottom, one bold line stood out: **Amount owed by patient: $1,240.00**.

"I cannot pay this," Carol whispered to the empty room. Her orange cat, Barnaby, jumped onto the chair next to her, letting out a soft, questioning meow.

Carol felt a cold knot of panic in her chest. She had trusted both her dentist and her insurance plan.

Now, she felt as if she faced a huge stone wall. The letter's language made her feel small, confused, and powerless.

Her first instinct was to call the insurance company and yell. She wanted to explain that she was a sixty-seven-year-old retired teacher on a fixed income and that their decision was unfair.

But she knew what would happen. She would get put on hold.

When she finally reached a person, they would repeat the confusing terms from the letter. She would get lost, her voice would shake, and she would end up hanging up in tears. Recognizing this, Carol knew that patience and persistence are key to navigating tough conversations.

She realized she had to understand their rules before she could argue her case. She needed someone to translate for her.

Carol glanced at her digital tablet on the counter. In the past few weeks, she had used her AI assistant to plan trips and write polite notes.

But this situation was different. It was a serious financial problem, and many might wonder if AI can reliably help with legal or insurance language. Clarifying this can build trust in AI as a practical tool for understanding complex documents.

She remembered the Three-Level Trust Test she had learned. This was a Level 3 case, a high-risk problem that affected her personal finances.

She knew she could not let the AI write a final letter and send it without checking. She needed to understand how to verify the AI's explanations to ensure accuracy before making decisions or communicating with the insurance company.

But she could use the AI as a patient, private tutor. It could help her break down the dense legal language, explain the hidden rules, and get her ready to talk to the insurance company. Trusting AI in this way can make complex problems feel more manageable and less intimidating.

Carol opened the app, mindful of privacy and data security. She took a deep breath and applied what she had learned about safety, context, and the importance of asking follow-up questions to ensure her sensitive information remained protected.

The Integration of Your AI Skills

To solve her dental crisis, Carol did not just type a single, hurried question. She did not behave like a search engine user.

Instead, she integrated three major skills she had built throughout her journey:

1. **The Safe Summary Method**: She read through the insurance letter and removed her policy number, her Social Security number, her dentist's real name, and her address. She kept the raw text of the legal clauses but protected her private identity.

2. **The Five-Part Prompt Formula**: She constructed a prompt with a clear Role (medical billing advocate), a specific Goal (explain the denial clauses and draft questions), rich Context (retired teacher on a fixed budget), a clean Format (a bulleted list with bold headings), and a strict Quality Check (no legal jargon in the explanation).

3. **The Conversation Partner Mindset**: She did not expect a perfect, finished solution on the first try. She prepared to ask follow-up questions, dig deeper into the insurance company's reasoning, and use the output as an educational stepping stone.

Here is the exact prompt Carol typed into her tablet:

> Act as a patient, highly detail-oriented medical billing advocate. I need your help to understand an insurance denial letter for a dental crown and a core buildup. I am a retired schoolteacher, and I want to understand their rules so I can call them and resolve this calmly.
> Here is our context: My dentist performed two procedures: a porcelain crown and a core buildup to support the crown. The insurance company denied both.
> Here is the exact text of their denial reason, which I have anonymized: 'Under our alternate benefit provision, the core

buildup is considered clinically unsubstantiated without radiographic verification of structural decay.'
Please explain this denial in plain, simple English that an absolute beginner can understand. Use a simple, real-world analogy to show what they are saying.
Finally, give me three polite, direct questions to ask the insurance representative when I call to appeal this decision.
Do not use any complex legal or medical jargon in your explanation."*

Carol pressed send. She watched the cursor blink on her screen and felt quietly determined.

Within seconds, the AI started to write. The response was warm, clear, and free of the cold, intimidating language from the original letter:

I understand how stressful it is to receive a large, unexpected medical bill. Let's make this simple and clear so you can resolve this.

The Denial Explained in Plain English: The insurance company is using a rule called 'alternate benefit.' This means they will only pay for the least expensive way to repair a tooth, unless the dentist proves that a more expensive treatment was necessary.

When they say the core buildup was 'clinically unsubstantiated without radiographic verification,' they are saying: 'We see your dentist built up the base of the tooth to hold the crown, but they did not send us an X-ray that clearly proves the tooth was too decayed to hold the crown on its own.'

A Simple Analogy: Imagine you hired a carpenter to fix a rotten wooden post on your front porch. The carpenter realizes the ground beneath is soft, so they pour a concrete base to prevent the porch from collapsing.

The insurance company is like an inspector who looks at the bill from a distance and says, 'We will pay for the wooden post, but we won't pay for the concrete base because you didn't send us a photograph proving the soil was actually

soft.'
Three Questions to Ask the Insurance Company:
1. Can you confirm if this claim was denied simply because you need the digital X-rays and clinical notes from my dentist showing the structural decay before the core buildup?
2. If my dentist submits a formal appeal with the preoperative X-rays and a narrative letter explaining why the tooth could not support a crown without a core buildup, will you re-evaluate this claim?
3. Is there a specific fax number or secure email portal where my dentist's office should send these supporting documents to expedite the review?

Carol read the screen and felt a wave of relief. The stone wall she had felt before was gone.

Now, it was just a simple, logical problem. The insurance company was not saying the procedure was wrong; they just wanted more paperwork. They needed the X-rays.

With this new understanding, Carol stayed calm. She did not raise her voice.

She called her dentist's office first. She spoke to the billing coordinator and read the second question from her screen.

The coordinator gasped. "Oh, Carol, I see what happened," she said. "We sent the claim for the crown, but our new assistant forgot to attach the digital X-rays showing the deep fracture. That is why they denied the buildup of the core. I will resubmit the files with the X-rays and a narrative letter today."

Two weeks later, Carol received a new statement. The insurance company had approved the claim. Her balance owed had dropped from $1,240.00 to a manageable $150.00 deductible.

She had won her battle not by letting the computer fight for her, but by using it to teach herself.

She had stayed in control of her own learning.

AI as a Plain-English Tutor

The real strength of artificial intelligence is not just writing quick emails or making nice images. Its greatest value is its endless patience as a personal tutor.

In a regular classroom, a teacher has to talk to thirty students at the same time. They teach at an average pace, using one textbook. If you do not get something right away, you have to ask in front of everyone or stay after class for help.

Your digital assistant, however, is a private tutor that belongs only to you.

It does not mind if you ask the same question ten times in different ways. It never gets tired, never judges you, and can draw from almost all of human history, science, and literature.

To make the AI an effective tutor, you need to move away from just listening to lectures. Instead, use the **Conversation Partner Mindset** to create an interactive, two-way learning process.

Customized Learning Levels

You can instruct the AI to adjust its vocabulary, tone, and complexity to match your exact level of knowledge on any topic.

Do not accept a basic, textbook answer.

You can ask the AI to explain something as if you are:

- A ten-year-old who loves soccer.
- A busy professional with only two minutes to read.
- A college student who needs to understand the reasons behind a historical event.
- A senior who wants a simple, everyday example.

The Plain-English Tutor Prompt

Use this template to begin learning about any difficult, confusing, or intimidating topic:

> Act as a warm, highly encouraging, and patient plain-English tutor. I want to understand [insert topic, e.g., how inflation works, the basics of blockchain, or how a car engine runs].
> I am an absolute beginner at this topic, and I find highly technical terms confusing. Please explain the core concept in three short paragraphs.
> In your explanation:
> 1. Avoid all technical jargon.
> 2. Use a simple, real-world analogy from daily life to illustrate the main point.
> 3. List the single most common mistake beginners make when trying to understand this.
> At the very end, ask me two simple multiple-choice questions to test my understanding, and wait for my response before giving me the answers.

Active Recall and Interactive Quizzing

When you read a book or watch a video on a new topic, your brain is mostly passive. It feels like you are learning, but you actually remember very little.

Real learning needs **active recall**. This means making your brain pull up information and answer questions. That mental effort is what builds real memory.

You can use your AI assistant to design custom quizzes, flashcards, and practice drills for any subject you want to master.

The Quiz Generator Prompt

If you are studying for an exam, learning a new language, or trying to understand a complex work policy, use this prompt to test your knowledge:

> Act as a supportive, expert study partner. I am trying to learn [insert topic].
> Please create a three-question multiple-choice quiz on this topic. Make the questions practical and focus on real-world scenarios rather than rote memorization.
> Deliver the questions one at a time. Do not show me the correct answers or the next question until I have typed my response to the current one.
> Once I answer, tell me if I was correct, explain why in one simple sentence, and then show me the next question.

This interactive approach turns studying from a boring task of reading old notes into an engaging game.

It makes you think, use what you know, and get instant, personal feedback without any stress or judgment.

Breaking Down Complicated Documents

In our daily lives, we are constantly forced to sign documents we do not understand. We sign apartment leases, software terms of service, home repair contracts, and employment agreements.

We see pages of dense legal text, our eyes glaze over, and we click "Agree" because we do not have the time or money to pay a lawyer to read it for us.

This is a big safety risk. You might agree to hidden fees, unfair cancellation rules, or give up your legal rights without knowing it.

You can use the **Safe Summary Method** to turn your AI assistant into a document translator.

It can scan these dense documents, point out hidden risks, and explain the terms in plain, simple language before you sign.

The Document Translator Prompt

Before you sign any contract, copy the text (making sure to remove real names, addresses, and financial account numbers) and run this prompt:

> Act as a meticulous, defensive consumer advocate. I am going to paste a section of a contract that I am preparing to sign. Please read through the text and organize your analysis into a clean, visual format with these three sections:
> **1. The Plain-English Summary**: Explain the primary purpose of this section in two simple, conversational sentences.
> **2. Hidden Risks and 'Red Flags'**: List any clauses, fees, penalties, or limitations that seem unfair, unusual, or risky for a consumer. Use bold text for the most critical risks.
> **3. My Safe Counter-Questions**: Provide three polite, clear questions I should ask the other party to clarify my rights before I sign this document.
> Here is the anonymized text of the contract: [paste the text here]

Let us look at a real-world example of how this works.

Imagine you are signing a contract for a new gym membership, and the text contains this clause: "Member agrees to indemnify and hold harmless the facility from any and all liability arising from accidental injury, regardless of negligence."

The AI will translate this dense legal phrase into a simple, clear warning:

> **Red Flag**: This clause is saying: 'If you get hurt while working out here, even if it is because our equipment is broken and we didn't fix it, you agree not to sue us or ask us to pay your medical bills.'
> **Your Counter-Question**: 'Could you please explain what your policy is if a member is injured due to a known mechanical failure of your equipment?'

Researching Safely Without Getting Fooled

When you use AI for research, you must remember the Confidence Trap. Because the AI is designed to predict words rather than verify facts, it can confidently invent fake facts, quotes, and historical dates that sound completely real.

To research safely and maintain control over the truth, you must use a hybrid approach.

You should use the AI to organize and guide your research, but rely on trusted, primary sources for the facts.

THE HYBRID RESEARCH CYCLE

1. USE AI TO: —> Brainstorm search terms, build outlines, and find major concepts.

2. USE GOOGLE TO: –> Find official primary documents, dates, and verified statistics.

3. USE AI TO: —> Summarize the long, verified documents and explain the confusing parts.

1. Let the AI Build Your Search Strings

If you are researching a complex topic, such as a medical condition, a historical event, or a new market trend, do not ask the AI for raw facts directly.

Instead, ask the AI to help you build highly effective search terms that you can type into traditional, trusted search engines like Google Scholar, official government websites, or public library databases.

Type this prompt:

> I am researching [insert topic]. I want to find high-quality, verified, and official primary sources (such as academic studies, government reports, or historical archives).

> Please suggest a list of five specific, highly targeted search strings (using advanced search operators like quotes or site-specific terms) that I can copy and paste into a search engine to find these documents.

2. Request Alternative Viewpoints

To avoid bias, force the AI to present the counter-arguments to any theory or plan you are researching.

Never accept a single, one-sided perspective. Type this follow-up:

> What is the most common, credible counter-argument or alternative perspective to the theory we just discussed?
> Please present both viewpoints side by side in a clean, two-column table. Focus on the main evidence used by each side, and do not declare one side as 'correct.'

Learning a New Skill with a 7-Day Sprint

Whether you want to learn how to play basic chords on a guitar, understand the fundamentals of watercolor painting, or learn how to use spreadsheet formulas at work, you can use your AI assistant to design a customized, step-by-step curriculum.

To avoid feeling overwhelmed, do not ask the AI for a full guide to a huge topic.

Instead, ask for a highly focused, short-term learning plan. This is what we call a **7-Day Learning Sprint**.

It breaks a big, intimidating skill into seven small daily exercises, each taking less than 15 minutes.

The 7-Day Sprint Prompt

Use this template to start building momentum on any skill you want to acquire:

> Act as an expert, highly encouraging coach who specializes in beginner-friendly training. I want to learn the fundamentals of [insert skill, e.g., basic Excel formulas, or physical stretching for flexibility].
> Here is our context: I am an absolute beginner, I have only 15 minutes a day to practice, and I want to avoid feeling overwhelmed or injured.
> Please design a custom, 7-Day Learning Sprint for me. Deliver the plan in a clean, day-by-day format. For each day, provide:
> **The Daily Focus**: A single, simple concept or movement to learn.
> **The 15-Minute Exercise**: A step-by-step practice drill.
> **The Common Pitfall**: One mistake to actively avoid today.
> **The Self-Check Question**: A simple question I can ask myself at the end of the day to make sure I did it right.

This structured plan takes away the stress of deciding what to study each day.

You do not have to spend hours searching for tutorials or videos. Just open your plan, do your fifteen minutes of practice, and build your confidence day by day.

Try It Now Exercise (5–10 Minutes)

Let us practice using the AI as an interactive plain-English tutor right now. This simple exercise will show you how quickly you can master a difficult topic through a guided conversation:

1. **Pick a Confusing Topic**: Think of a complex term or concept that you have always found confusing (such as how inflation works, the difference between stock options and regular stocks, or how GPS technology knows where you are).

2. **Run the Tutor Prompt**: Open your AI tool, and paste this conversational instruction: > "Act as a patient, encouraging plain-English tutor. I want to understand [insert your chosen

topic here] in simple terms. Please explain the core concept in three short paragraphs, using a warm, everyday analogy. Do not use any technical jargon. > > At the very end of your response, ask me a single multiple-choice question to test my understanding, and wait for my answer before replying."

3. **Answer the Question**: Read the explanation and the AI's question. Type your answer (e.g., "A" or "B") into the box.

4. **Read the Feedback**: See how the AI checks your answer, explains why you were right or wrong, and helps you understand more, all without stress or embarrassment.

Prompt Rescue Box

The Problem

The AI's explanation of a complex topic is still too dense, academic, or technical to follow. It uses words that make you look up more definitions, leaving you even more confused.

The Rescue Prompt

Do not start a new chat. Paste this exact message into your existing conversation:

> That explanation was still too technical and difficult for me to follow. Let's pause and simplify.
> Imagine you are talking to a ten-year-old child who has never studied this topic before.
> Please rewrite your explanation. Use only very simple, everyday words and short sentences.
> Use a simple, physical analogy that I can picture in my mind, and keep the whole explanation under five sentences."

Human-in-Charge Check

Always remember the core truth of modern technology: **AI is an excellent assistant for summarizing and explaining, but real learning requires human effort**.

It is very easy to let the AI read, summarize, and outline everything for you. You might feel smarter because your screen is full of neat bullet points and summaries.

But if you cannot explain the idea in your own words to a friend, without looking at your screen, you have not really learned it.

The AI is your assistant, not your replacement. Let it help clear the way, but make sure you are the one doing the real thinking.

Action Steps & Checklist: Learning With AI

Use this quick checklist to review your learning and research interactions, ensuring you are building real, lasting understanding:

☐ **Did I anonymize my documents before pasting them?** (Have I removed all real names, account numbers, and private details using the Safe Summary Method?)
☐ **Did I instruct the AI to use simple, plain English?** (Did I explicitly ban complex technical jargon and academic transition words?)
☐ **Did the AI explain the concept using a clear, real-world analogy?** (Can I easily picture the physical example in my mind?)
☐ **Have I tested my understanding using interactive questions?** (Did I ask the AI to quiz me rather than just reading its summaries passively?)
☐ **Did I fact-check the key dates, names, and statistics?** (Have I verified high-risk claims against a trusted, primary database?)
☐ **Can I comfortably explain this topic to someone else in my own words?** (Can I speak the truth of the concept without relying on the computer's script?)

Carol Higgins stood on her back porch in Sarasota, holding a small green watering can. The afternoon sun finally broke through the gray clouds, shining warmly on the damp ferns hanging from the rafters. Barnaby sat on the wooden deck by her feet, eyes closed, enjoying a patch of sunlight.

Carol felt a quiet, deep sense of triumph.

The heavy manila envelope with the scary dental bill still sat on her kitchen table. But now, it was no longer a symbol of fear or confusion. It was just a piece of paper, a problem she had faced, understood, and solved.

She had not let her anxiety stop her. She had not let the dense, scary legal language force her to pay money she did not owe.

She had stayed in control. She used the Safe Summary Method to protect her privacy and guided her digital assistant step by step until the assistant clearly understood the rules.

She made the final, personal connection with her doctor's office and won her battle. She was, without a doubt, in charge of her digital future.

As she watered her plants, Carol realized her whole view of technology had changed.

A few weeks ago, she saw the empty chat box as something intimidating, a fast, scary tool meant only for programmers and tech experts.

Now, she saw it as a place to work together. It was a patient tutor, a helpful writing assistant, and a quiet administrative helper she could control.

She had used it to plan trips, write letters, understand bills, and organize her life. She felt more confident, more organized, and more connected to the world than she had in years.

But as she put down her watering can, a new question came to her mind.

It was great that she had found ways to use AI for these one-time tasks. The templates and formulas were very helpful.

But thinking about her busy daily routine, she realized she only opened the app when a problem came up. She was still treating her assistant like someone who only helps in emergencies, not as a regular part of her day.

How could she move from using AI for random, one-time questions to making it part of her daily routine?

How could she create a simple daily habit and a personal workspace to help her manage her calendar, plan meals, organize her hobbies, and save hours of stress each day?

To do that, she would need more than just a few formulas. She would need a complete, personal workflow and a simple, step-by-step challenge to build lasting digital confidence.

Chapter Ten

Build Your Personal AI Workflow

Back in 1911, people in construction believed bricklayers could only speed up by working harder. For centuries, the process stayed the same: a worker would bend down, pick up a heavy brick, check the best side, mix mortar, set the brick in place, and clean off the excess mortar.

Most people thought some workers were just naturally faster, while others were slower. If a builder needed to finish a wall quickly, the only option was to hire more workers and push them to work harder.

Frank Gilbreth, an industrial engineer, challenged this idea. He didn't search for workers with special skills or force them to work faster. Instead, he quietly observed bricklayers with a camera and a notebook, paying close attention to every move they made.

Gilbreth found that a typical worker made eighteen different movements for each brick. Most of these steps wasted energy. Workers had to bend down, pick up a heavy brick, lift it, and turn it around. All this bending and lifting wore them out before lunchtime.

Instead of focusing on how fast each worker was, Gilbreth looked at the whole system. He built an adjustable wooden scaffold that could be raised as the wall got higher. This kept the bricks at waist level, so workers didn't have to bend down anymore.

He also hired assistants to sort the bricks in advance, placing them on trays with the best side facing up. He even developed a mortar recipe that was always the right consistency, so bricklayers only needed to tap each brick once rather than several times.

With these changes, Gilbreth reduced the number of movements from eighteen to fewer than five. The job became much less tiring.

Bricklayers didn't have to work harder or sweat more, but their output jumped from 120 to 350 bricks per hour. This wasn't because of a secret trick; it was because they followed a repeatable system.

Today, many of us make the same mistake at our computers that builders made back in 1911.

We see people online selling huge spreadsheets of so-called 'magic prompts.' We save websites, store complicated templates, and try to collect every secret shortcut. We think that if we find the perfect phrase, our work will become easy.

This is the trap of collecting prompt lists. Gathering thousands of prompts doesn't make you faster or more confident. It just clutters your computer and adds a new kind of mental fatigue.

Prompts can become outdated as soon as a tech company updates its software. Even worse, searching through a huge list of other people's templates often takes longer than just writing your own message.

You don't need a huge collection of prompts. What you really need is a simple, repeatable system. It's time to stop collecting prompts and start building workflows.

The Prompt Collector vs. The Workflow Builder

To understand why this shift matters, we can look at two different managers working for the same training company.

Mark is a Project Operations Manager in Chicago. He is a prompt collector. He has a digital folder on his computer desktop called "My AI Power Tools." Inside the folder are three Word documents, five bookmarked websites, and a spreadsheet containing over 4,000 prompts he bought from an online fitness and business creator.

On Monday morning, Mark needs to write a simple project update email to his team. He opens his spreadsheet and starts scrolling through it. He looks through "Email Prompts," then "Business Communication," and finally "Project Management."

He finds a template that looks promising. It is eighty words long, filled with complex instructions and empty brackets where he is supposed to type his own details.

Mark spends fifteen minutes copying the template, typing his project details into the brackets, and running it through the AI.

The output is a long, stiff message that doesn't fit his team's real relationships. He has to spend another ten minutes editing the text by hand on his screen. The entire process takes him nearly half an hour, leaving him frustrated and behind schedule.

Paula Jenkins, working in her Raleigh home office, approaches her Monday morning in a completely different way. She does not have a spreadsheet with 4,000 prompts. She does not collect templates.

Instead, she has a single, three-step routine that she completes in less than ten minutes.

Paula has a folder on her computer's bookmark bar with just three saved links. Each link is a custom workflow she designed herself using the Five-Part Prompt Formula. She knows exactly when to use them, what information to feed them, and how to verify the results.

Every Monday at 8:00 a.m., her morning workflow begins with a simple physical trigger: she opens her calendar.

She has a saved prompt that turns her messy weekly task list into a clean, prioritized schedule. She pastes her raw notes into the box, runs her prompt, double-checks the dates, and copies the clean schedule into her team's shared workspace.

She does not waste time searching for secret keys. She runs her system, finishes her work, and enjoys her morning coffee.

The Five-Part Workflow

A workflow is not a magical phrase. It is a repeatable path that turns raw, messy information into a polished, usable, real-world result.

Every successful workflow you build with your AI assistant must contain five distinct parts:

TRIGGER —> TASK —> PROMPT —> REVIEW —> ACTION

1. **The Trigger**: This is the physical event, time, or feeling that tells you it is time to use AI. It answers the question: When do I start? (e.g., "Every Friday at 3:00 p.m.," or "When my email inbox has more than ten unread messages").

2. **The Task**: This is the specific, administrative chore you want to delegate to your assistant. It answers the question: What exact job am I trying to complete? (e.g., "Turn my rough meeting notes into a clean bulleted list").

3. **The Prompt**: This is your structured, five-part instruction template. It answers the question: How do I safely tell my assistant what to do?

4. **The Review**: This is your defensive check for safety, tone, and accuracy. It answers the question: How do I make sure the assistant did not make a confident mistake or share private data?

5. **The Action**: This is the physical step you take with the finalized text. It answers the question: Where does this work go now? (e.g., "Send the email," "Print the packing list," or "Paste the schedule into the team channel").

By organizing your daily tasks into these five steps, you remove all the friction of using AI. You no longer have to stand in front of the empty chat box, wondering what to type. Your system is already built, waiting for your raw materials.

Five Everyday Workflows

To help you get started, here are five practical, everyday workflows that you can build and use today. Each one is designed to save you hours of stress and mental fatigue at work, in your writing, in your learning, and in your home life.

1. The Weekly Planning Workflow

Use this system every Sunday evening or Monday morning to turn a chaotic, overwhelming mental list of chores into a clean, peaceful schedule.

- **Trigger**: Every Monday at 8:00 a.m., before you open your email inbox.
- **Task**: Organize a messy, stream-of-consciousness list of weekly tasks into a prioritized, day-by-day calendar.
- **The Prompt**:

> Act as an organized personal assistant. I am going to provide a raw, messy list of my tasks, appointments, and chores for the upcoming week. Please organize these tasks into a clean, day-by-day schedule from Monday through Friday.
> For each day, list:
> - **Top Priority**: The single most important task to complete first.
> - **Secondary Tasks**: Up to three smaller chores to work on later.
> - **Quick Wins**: Any quick, five-minute tasks that can be completed between meetings.
> Keep the language simple, direct, and practical. Here is my raw task list: [Paste your messy list here].

- **The Review**: Make sure the AI did not schedule two major appointments at the same time, and verify that the deadlines match your real calendar.
- **The Action**: Copy the organized schedule into your physical planner, your calendar app, or a clean document on your computer.

2. The Email Reply Workflow

Use this system whenever you receive a difficult, long, or emotionally draining email that requires a delicate, professional reply.

- **Trigger**: When you open an email that makes you feel anxious, annoyed, or unsure of how to respond.

- **Task**: Draft a polite, clear, and collaborative response that protects your boundaries.

- **The Prompt**:

> Act as a professional communication coach. I need your help to draft a reply to an email.
> Here is our context: I want to reply to [describe the sender, e.g., a client, my manager, a neighbor]. My goal is to [describe your goal, e.g., decline their invitation politely, explain why a project is delayed, or ask for clarification on a bill]. The tone of my response should be [e.g., warm but firm, highly professional and brief].
> Please deliver the draft using short, one-sentence paragraphs, so it is easy to read on a phone. Keep the entire response under four sentences, and do not use any defensive apologies or corporate clichés.
> Here is the email I received: [Paste the anonymized text of the email here].

- **The Review**: Read the draft out loud. Make sure it doesn't sound defensive or robotic, and make sure you have removed all real-world names or private details.

- **The Action**: Copy the text, paste it into your email client, manually type the recipient's real name, and hit send.

3. The Fridge-to-Dinner Workflow

Use this system on busy weeknights when you are tired and hungry and do not want to spend money on expensive takeout.

- **Trigger**: Every Tuesday and Thursday at 6:00 p.m., when you stand in front of your open refrigerator wondering what to cook.

- **Task**: Generate a simple, quick recipe using only the random ingredients you already have in your kitchen.

- **The Prompt**:

> Act as a practical, creative home cook. I want to make a simple dinner tonight using only the ingredients I have on hand. I want to avoid going to the grocery store.
> Here is my context: I have [list your ingredients, e.g., two chicken breasts, a half-used onion, a can of black beans, some white rice, and basic spices]. I only have twenty minutes to cook, and I want to minimize the number of dirty dishes.
> Please suggest a single, simple recipe I can make tonight.
> Provide:
> - **The Recipe Name**: Something simple and appetizing.
> - **Prep Time and Cook Time**: Estimated in minutes.
> - **Step-by-Step Instructions**: Clear, numbered steps.
> - **The Cleanup Tip**: How to make washing up easier.
>
> Do not suggest any ingredients that I did not list above.

- **The Review**: Check the steps to make sure you actually have the required tools (such as a frying pan or a baking sheet) and make sure the cooking times seem safe and realistic for meat.

- **The Action**: Set your tablet or phone on the kitchen counter and start cooking.

4. The Travel Packing Workflow

Use this system three days before any trip to ensure you travel light, stay comfortable, and avoid the stress of overpacking.

- **Trigger**: Three days before your departure date, when you pull your suitcase out of the closet.
- **Task**: Create a simplified, layer-based packing checklist suited to your destination's weather and your physical limits.
- **The Prompt**:

> Act as a practical, experienced travel planner. I am preparing for a [number-of-days trip, e.g., five-day] trip to [destination] in [month]. Here is my context: I am [your age/physical preference, e.g., 67 years old] and want to travel with only one light carry-on bag. The weather forecast is [e.g., sunny afternoons around sixty degrees, but chilly nights around thirty degrees]. We will be doing [e.g., gentle walking tours and backyard dinners].
> Please create a simplified packing checklist. Focus on items that can be easily layered to save space. Deliver the list in a clean, bulleted format, grouped by category: **Clothing (Layers)**, **Shoes**, **Toiletries**, and **Travel Documents**. Keep the list under fifteen total items.

- **The Review**: Make sure the list does not contain any unnecessary heavy gear or duplicate items.
- **The Action**: Print the list or keep it open on your phone as you place your clothes into your suitcase.

5. The Meeting Prep Workflow

Use this system 30 minutes before any important video call, parent-teacher conference, or doctor's appointment to stay organized and get your questions answered.

- **Trigger**: Thirty minutes before a scheduled high-stakes meeting or appointment.
- **Task**: Generate a structured agenda and a list of three specific, polite questions to guide the conversation.
- **The Prompt**:

> Act as an organized professional advocate. I am preparing for a meeting with [describe the person, e.g., my child's teacher, my doctor, a new client] to discuss [describe the topic, e.g., my child's progress in reading, a sudden knee pain, a project rollout]. Here is our context: The meeting is scheduled for [e.g., 20 minutes]. My primary goal is to [state your goal, e.g., find out how to support my child's reading at home, or get a clear plan for my knee recovery].
> Please deliver a simple, one-page meeting guide.
> Provide:
> - **A 30-Second Opening Statement**: A polite, clear sentence to start the meeting and state my goal.
> - **Three Polite, Direct Questions**: Focused on getting actionable answers.
> - **The Next Step Reminder**: A question to ask at the very end to ensure we are on the same page.
> Keep the language simple, direct, and warm.

- **The Review**: Read the opening statement out loud to ensure it sounds natural to your voice and fits the relationship.
- **The Action**: Write the questions down in your notebook or keep

them open on your screen during the call.

The 7-Day AI Confidence Challenge

Building a new habit can feel intimidating, especially when you are busy. To help you transition from reading about these tools to using them with complete confidence, we have designed a simple, day-by-day practice plan.

This is the 7-Day AI Confidence Challenge.

Each day, you will spend less than ten minutes completing one small, low-risk task. You will not have to write complex code or risk any private data.

You will practice your conversational skills, test your formulas, and build lasting digital habits.

7-DAY CONFIDENCE CHALLENGE

Day 1: Clear Question
Day 2: Plain Tutor
Day 3: Voice Edit
Day 4: Org Notes
Day 5: Simple Plan
Day 6: Prompt Rescue
Day 7: Build Your Personal Workflow

Day 1: Ask One Clear Question

- **The Goal**: Experience a successful, specific response using your very first conversational prompt.
- **What to Do**: Open any free AI tool. Think of a simple, everyday task you need to complete today (such as writing a polite text message to decline a dinner invitation, or finding three quick ways to clean a coffee stain).

- **The Prompt**:

Act as a friendly, patient assistant. I am new to using AI, and I want your help to draft a quick, polite text message declining a neighbor's dinner invitation because I am tired. Keep the message under two sentences, and make sure the tone sounds warm and friendly.

- **The Self-Check**: Did the AI deliver exactly what you asked for without any long-winded introductions?

Day 2: Ask AI to Explain Something

- **The Goal**: Use the AI as a patient, plain-English tutor to master a complex topic.
- **What to Do**: Think of a technical term, historical event, or concept that you have always found confusing (such as how a cloud stores data, how credit card interest works, or the difference between baking soda and baking powder).
- **The Prompt**:

Act as a patient, encouraging plain-English tutor. Please explain [insert your topic here] to me as if I were a ten-year-old beginner. Use a simple, real-world analogy that I can picture in my mind, and keep the explanation under three short paragraphs.
At the very end, ask me one simple multiple-choice question to test my understanding, and wait for my response.

- **The Self-Check**: Read the analogy. Can you now comfortably explain the concept to a friend in your own words? Reply to the AI's question and see how it checks your work.

Day 3: Improve an Email or Message

- **The Goal**: Use the Voice Calibration System to train the AI on your natural writing style.
- **What to Do**: Find a short email or message you wrote in the past that sounded warm, natural, and authentic to your voice.
- **The Prompt**:

Act as a professional writing coach. I want to train you on my personal writing voice. Below, I have pasted a sample of my past writing.
Please analyze my style in three simple bullet points, focusing on my sentence length, my vocabulary level, and my tone.
Here is my sample: [Paste your short, anonymized sample here].

- **The Self-Check**: Does the AI's analysis match how you actually sound in real life? Save this style analysis in your phone notes.

Day 4: Summarize or Organize Notes

- **The Goal**: Turn administrative clutter into a clean, visual structure.
- **What to Do**: Take a messy, disjointed block of notes from a recent project, a meeting, or a personal list of home repair tasks.
- **The Prompt**:

Act as an organized administrative assistant. Please take my messy notes and organize them into a clean, professional

summary.
Give me the output in this format: use a short, two-sentence executive summary at the top, followed by a simple bulleted list of key action items, and use bold text for the most urgent deadlines.
Here are my raw notes: [Paste your messy notes here].

- **The Self-Check**: Is the organized summary easier to read and use than your original block of notes?

Day 5: Plan Something Useful

- **The Goal**: Use the Five-Part Formula to plan a real-world task without feeling overwhelmed.
- **What to Do**: Think of a small project you need to complete soon (such as cleaning out your garage, preparing a presentation for work, or organizing a weekend family barbecue).
- **The Prompt**:

Act as a practical planning assistant. I want to plan [describe your event/project]. Here is my context: I have [e.g., two hours this Saturday, a budget of $50, or three volunteers helping me].
Please make a simple, step-by-step checklist running in chronological order. For each step, list the single tool or item I will need. Keep the plan under seven total steps.

- **The Self-Check**: Did the AI stay within your strict budget and timeline constraints?

Day 6: Rescue a Bad Answer

- **The Goal**: Practice using the Prompt Rescue System to diagnose and fix a failing conversation on the fly.
- **What to Do**: Intentionally run a vague, bad prompt that is guaranteed to produce a robotic, cliché-filled response: "Write a sales pitch for a business."
- **The Prompt (The Rescue)**: Once the AI delivers its stiff, high-pressure response, type this exact rescue script into the same chat window:

This response is far too sales-heavy and generic. Let us rescue this draft. Please rewrite this pitch using simpler, direct English and shorter sentences.
Focus on a warm, supportive, and neighborly tone. Do not use any exclamation marks, marketing hype, or clichés like 'unleash,' 'embark,' or 'testament.' Keep it under 100 words.

- **The Self-Check**: Notice how quickly the assistant abandoned the robotic tone and delivered a clean, human-centered draft once you set the standards.

Day 7: Build Your Personal AI Workflow

- **The Goal**: Create your first repeatable, custom workflow for a task you perform regularly.
- **What to Do**: Look back through your week. Identify one recurring administrative chore, writing task, or learning need that causes you stress or wastes your time.
- **The Action**: Use the Master Prompt Template to write your own

custom prompt. Save it in your personal library, and schedule your first Trigger on your calendar.

Your Personal Prompt Library

Once you have completed the challenge, you will have a small, highly effective collection of prompts that consistently deliver great results for your life and work.

Do not hide these prompts in a giant, confusing spreadsheet. Keep them simple, accessible, and safe.

- **Use a Simple Phone Note**: Create a single note on your phone or computer called "My AI Assistant."
- **Organize by Category**: Use clean headings to group your prompts (e.g., Work, Writing, Home, Learning).
- **Add Safety Reminders**: Write your digital boundaries directly at the top of your prompt note in bold letters: "**NEVER PASTE PASSWORDS, REAL NAMES, OR COMPANY FINANCIALS**." This is a permanent reminder before you copy any text.
- **Keep It Living**: As your daily tasks change or your software updates, feel free to tweak your prompts. If a follow-up question works particularly well for fixing tone, add it directly to your saved template.

Staying Current Without Feeling Overwhelmed

Every single week, there is a new headline about artificial intelligence. A technology company announces a faster model, a new phone app is released, or an online guru claims that a brand-new feature will "change everything forever."

It is incredibly easy to feel a quiet, constant pressure to keep up. We worry that if we do not spend hours reading tech blogs or learning new software buttons, we will eventually be left behind.

This pressure stems from a misunderstanding of how technology works.

Software interfaces, menus, and brand names will always change. Today you might use one tool; tomorrow your company might switch to another.

If you focus on learning which buttons to click in a specific app, you are building your house on shifting sand. You will constantly feel stressed, confused, and out of date.

But **communication principles** never change.

The art of asking a clear, specific question, sharing rich human context safely, and guiding a conversation politely is a permanent skill. It works on every model, in every app, and in every language.

If you master being the human in charge, you will never have to worry about a software update again. No matter what new tool is released in the future, it will still contain an empty prompt box.

And you will always know exactly how to talk to it.

Try It Now Exercise (5–10 Minutes)

Let us take your very first step in the 7-Day Confidence Challenge right now. Open your preferred AI tool, and follow these simple steps to build your daily momentum:

- **Copy the Template**: Highlight and copy the following starter prompt:

> Act as a patient, encouraging AI coach. I am new to using AI, and I am completing Day 1 of my confidence challenge. My goal is to write a warm, two-sentence text message to my friend, Sarah, thanking her for hosting a lovely dinner last night.
> Please deliver the draft in a simple, neighborly tone. Do not use any formal business language or emojis."

- **Run the Prompt**: Paste the text into the empty chat box and press send.

- **Edit the Draft**: Read the response. Spend ten seconds making any minor adjustments to ensure the text matches your natural voice.

- **Schedule Day 2**: Open your calendar or your task manager, and write down: "Day 2 AI Challenge: Explain a concept" for tomorrow at 9:00 a.m.

Workbook Element: My Three AI Workflows

Use this simple, physical template to design your personal routines. Pick three tasks in your life or work that you want to simplify, and write down their structures:

Workflow 1:

- **My physical trigger is:** _______________
  - **The specific task is:** _______________
  - **My saved prompt template is:** _______________
  - **My safety and review step is:** _______________
  - **My final physical action is:** _______________

Workflow 2:

- **My physical trigger is:** _______________
  - **The specific task is:** _______________
  - **My saved prompt template is:** _______________
  - **My safety and review step is:** _______________
  - **My final physical action is:** _______________

Workflow 3:

- **My physical trigger is:** _______________
  - **The specific task is:** _______________
  - **My saved prompt template is:** _______________
  - **My safety and review step is:** _______________
  - **My final physical action is:** _______________

When you build these workflows, you move away from the chaos of technology and create a calm, organized system that works for you. Instead of just collecting shortcuts, you become like Frank Gilbreth, setting up a system that makes your daily work easier and less tiring.

You're no longer just watching technology from the sidelines. Now, you're the one building, designing, and confidently managing your own productivity.

As you finish this chapter and get ready to start the 7-Day Challenge, look ahead with confidence.

Throughout this book, you've gained a strong set of skills. You know how to use AI as a conversation partner, apply the Five-Part Prompt Formula, protect your privacy, fix bad answers, and create repeatable workflows.

But how do these skills become part of your long-term mindset? How can we move beyond daily tasks to see the bigger picture in our lives, communities, and future?

How can we make a lasting commitment to stay in control as humans, even as the world keeps speeding up?

Conclusion

The afternoon sun lit up a golden stripe on the wooden floor of Paula Jenkins's home office in Raleigh. It was 4:30 p.m. on a Friday, usually the most stressful hour of her week. Most Fridays, Paula faced a pile of unfinished emails, messy meeting notes, and a sore neck from hours at her desk.

But today was different. Her desk was tidy, her mind was calm, her weekly reports were done, and her inbox was empty.

She spent the last 30 minutes using her weekly planning routine, turning a messy list of tasks into a clear, prioritized schedule for next week. She did all this safely, never sharing any private company information.

Feeling quietly satisfied, Paula closed her laptop. She stood up, stretched, and headed to the kitchen to start the weekend early with her family. Her success was not a miracle. She had just learned to manage her digital assistant with real confidence.

In East Nashville, Tennessee, the doors of Iron & Oak let in the cool evening breeze. Sam Vance stood on the gym floor, watching his members laugh and talk as they packed their bags.

The gym was thriving, and his latest email newsletter, which celebrated Maria's victory over her joint pain, really resonated with the community. That morning, five new parents signed up for his strength class, saying his newsletter felt warm, honest, and personal.

Sam wrote that newsletter in less than fifteen minutes at his kitchen table. He avoided marketing hype and robotic language. Instead, he guided the AI to match his natural, supportive voice and kept his clients' private details safe. By saving hours of writing time, he could focus on what he loved most, helping people get strong at the gym.

In Sarasota, Florida, Carol Higgins sat on her back porch with a warm mug of tea. Her orange tabby cat, Barnaby, purred on the chair next to her.

Carol's tablet rested on her lap. She used it to explore a new hobby, searching for the best ways to grow organic heirloom tomatoes in Florida's sandy soil.

She was no longer afraid of the empty chat box. She stopped treating it like a search engine and didn't panic when it gave her a long, confusing list of steps.

She learned to use the "Interviewer Prompt" to make the AI act like a patient gardening coach. She could ask for simple analogies, spot mistakes, and check information using her local library's website. Technology was no longer a fast-moving storm. Now, it was her own quiet workspace.

Paula, Sam, and Carol didn't go back to school to study computer science. They didn't learn to code or spend lots of money on secret prompt databases.

By understanding common misconceptions, like treating the blinking cursor as a test or fearing the chat box, you can avoid mistakes and build confidence in your ability to communicate effectively with AI.

How the Pieces Fit Together

Throughout this book, you have built a strong set of communication skills. It might seem like these tools are just separate tricks for different tasks, but they all fit together as part of one bigger pattern.

The more you understand how to adapt your prompts for various professional or personal contexts, the more effectively you can customize interactions, making your communication with AI more relevant and efficient.

The **Five-Part Prompt Formula**, Goal, Context, Role, Format, and Quality Check, gives you the tools to shape the assistant's first response. It helps you feel capable and confident in guiding AI, making your interactions more effective.

The **Safe Summary Method** protects you, so you can give the assistant helpful context without risking your private information. This reassurance helps you feel secure and in control of your data while still getting tailored results.

The **Follow-Up Loop** and the **Prompt Rescue System** help you stay in control. They remind you that the first answer is just a rough draft. Using the seven-step diagnostic system fosters resilience and confidence as you refine your results.

Finally, the **Three-Level Trust Test** makes sure you stay in charge of what's true. The machine can analyze and suggest, but only you are responsible for the final result.

When you use these skills in your daily routines, you stop feeling anxious about technology. Instead, you start shaping your own digital workspace and take control of your tasks.

Your New AI Confidence

You are no longer a tech-anxious beginner, confused by prompts and intimidated by the empty chat box.

You don't need to memorize thousands of prompt templates. You also don't need to worry about every small software update or new app release.

Menus and buttons will change, but clear, safe, and goal-focused communication is a skill you'll use forever.

Now you have a reliable way to communicate that works for any task, in any tool, for the rest of your life.

You understand the central truth of the digital age: **your own intelligence remains the most important part of the process**.

AI can write, suggest, draft, and explain. But only you can decide what matters, what is true, what is kind, and what is useful. You are in charge, and the assistant follows your lead.

The Human-in-Charge Commitment

To keep your confidence and protect your unique voice in a fast-changing digital world, commit to these six core principles:

- **Always state a clear, outcome-focused goal**: Never use vague, passive language. Tell your digital assistant precisely what physical action and scope you want to achieve.

- **Protect your private data at all times**: Treat the prompt box like a public billboard. Never paste passwords, Social Security numbers, financial ledgers, or real names into a public tool.

- **Treat the first response as a working draft**: Never copy and paste the AI's first output mindlessly. Always use the follow-up loop to shape, shorten, and refine the text.

- **Apply the Three-Level Trust Test**: Be highly skeptical of any statistics, quotes, or academic citations the AI provides. Verify all critical claims against trusted primary databases.

- **Read your drafts out loud**: Actively remove robotic clichés and corporate transition words. Ensure the final writing matches your natural conversational rhythm and authentic voice.

- **Maintain final human responsibility**: Never let the machine make high-stakes medical, financial, legal, or emotional decisions for you. Always use your own critical thinking and professional judgment to guide the final action.

Your First Action Steps Tomorrow Morning

Your journey as a confident, safe, and efficient prompter doesn't end with this book. It starts with your next digital interaction.

To make your new skills a habit, try these three simple steps tomorrow morning:

1. **Build Your First Workflow**: Select one simple, recurring task that causes you daily stress (such as taming your morning inbox or planning your family dinners). Write down its five-part workflow structure in your notebook.

2. **Save Your Master Template**: Copy the Five-Part Prompt Template from this book and save it to your phone or computer notes. Write your privacy boundaries in bold letters at the very top.

3. **Teach the Formula**: Share the Five-Part Prompt Formula with one friend, family member, or colleague who is feeling overwhelmed by technology. Teaching others is the absolute best way to lock in your own mastery.

The empty chat box is no longer something to fear. It's a blank canvas waiting for your ideas, your values, and your voice. Approach it with confidence.

Open your screen, take a deep breath, and start your next conversation. Remember, you are in charge. Keep that in mind with every prompt you write.

A Personal Request from the Author

Thank you for reading *How to Ask Artificial Intelligence for Better Results: A Human-in-Charge AI Prompt Guide for Beginners to Fix Weak Answers, Protect Privacy, Save Time, and Use AI with Confidence.*

I hope this book helped you see AI as a useful tool you can guide with confidence, instead of something confusing or intimidating. Whether you tried the Five-Part Prompt Formula, the Prompt Rescue System, the Human-In-Charge Checks, or the practical examples, I hope you now feel ready to ask better questions and get better answers.

Others could benefit from this book, too, and you can help them.

Honest reviews help others decide if a book is right for them. Your review doesn't have to be long or perfect. Just a few genuine sentences about what you found helpful or encouraging can make a big difference.

If this book helped you understand AI, write better prompts, avoid mistakes, protect your privacy, or feel more confident, please consider leaving an honest review.

Your review could help someone else feel less overwhelmed and start using AI safely and confidently.

Thank you for your time, trust, and support.

— George Munson

Quick Prompts

Bonus 1: Quick Reference Prompt Formula

Looking at an empty chat window can make you feel stuck, leading to unclear prompts and generic answers. The **Five-Part Prompt Formula** helps by turning that blank screen into a focused, useful conversation. Rather than memorizing lots of outdated prompts, keep this simple framework handy to create strong prompts whenever you need them.

"Goal"+"Context"+"Role"+"Format"+"Quality Check"

1. Goal (The What)

This is the base of your request. Clearly state the exact result you want. Swap out vague phrases like "help me write" or "tell me about" for clear, action-focused instructions. For example, say "Draft a polite email asking for a late report" instead of just "Email coworker."

2. Context (The Why and Who)

Share the background details that explain the situation. Mention your audience, any limits, style choices, or important details. Always remove any personal or company information to keep your data safe. For example: "The reader is a busy client who values directness, and we have a strict project deadline on Friday."

3. Role (The Perspective)

Giving the AI a role helps it choose the right words, tone, and expertise. Pick realistic, job-focused roles instead of over-the-top titles. For example: "Act as a patient writing coach" or "Act as an organized project operations manager."

4. Format (The Structure)

Say exactly how you want the answer to look. This makes sure the output is organized, easy to read, and ready to use right away. For example: a two-column comparison table, a step-by-step checklist, or a three-bullet summary.

5. Quality Check (The Guardrails)

Set clear rules to avoid robotic phrases, spot assumptions, or highlight accuracy risks. This step helps the AI check its answer before giving it to you. For example: "Avoid corporate jargon and marketing hype. List any factual claims that require external verification."

The Master Prompt Template

Copy, paste, and adjust this main prompt for any daily task:

> Act as a [Role]. I want to **[Goal]**. Here is the context: **[Context]**. Please give me the output in this format: **[Format]**. Before you finish, make sure to check for **[Quality Check]**.

Just fill in the brackets to skip trial and error and get results that fit your needs right away. Use this template whenever you want a quick way to start.

Bonus 2: Prompt Rescue Cheat Sheet

If your chat session starts to go off track, you probably do not have time to reread instructions or start over. You need a quick, copy-and-paste rescue prompt to get the AI back on course.

Bookmark this cheat sheet. If an output does not meet your needs, find the problem below, copy the matching rescue prompt, fill in the blanks, and paste it into your chat.

1. The Output Is Too Vague

The Diagnostic: The AI gave broad, hand-waving suggestions or high-level concepts that lack practical steps.

The Rescue Prompt:

> Make this response more specific for [insert your specific audience, situation, or goal]. Replace the general advice with three concrete, real-world examples, and list the exact steps I need to take next.

2. The Output Is Too Long

The Diagnostic: You asked for a quick summary or a brief email, but the AI returned an overwhelming wall of text.

The Rescue Prompt:

> This is too long and detailed for what I need. Cut this response by 50%. Keep only the most critical points, organize them into a clean bulleted list, and put the single most important action step in bold.

3. The Output Is Too Short

The Diagnostic: The AI gave a lazy, brief answer that lacks the necessary context, depth, or instructions.

The Rescue Prompt:

> This answer is too brief. Expand on this draft by adding practical step-by-step instructions, explaining the underlying reasoning, and listing two common pitfalls I should actively avoid.

4. The Output Sounds Too Robotic

The Diagnostic: The text sounds stiff, clinical, or like an instruction manual. It lacks warmth and human rhythm.

The Rescue Prompt:

> Rewrite this so it sounds more natural, warm, and human. Use plain language, varied sentence lengths, and a friendly but professional tone. Avoid stiff transitions and passive verbs.

5. The Output Is Too Technical

The Diagnostic: The AI used heavy industry jargon, confusing acronyms, and complex theories, making the explanation hard to follow.

The Rescue Prompt:

> That explanation was too technical. Explain it again as if you are talking to a beginner who has never studied this topic. Use a simple, real-world analogy to make the core concept clear.

6. The Output Is Too Generic

The Diagnostic: The text is stuffed with tired marketing slogans and clichés, such as "delve," "in today's fast-paced world," "testament," and "embark on a journey."

The Rescue Prompt:

> Rewrite this draft to remove all marketing clichés, hype, and generic transition words (like 'delve,' 'moreover,' and 'in conclusion'). Use simple, direct, conversational language and focus on realistic facts.

7. The Output Is Factually Suspicious

The Diagnostic: The AI cited a questionable statistic, an unverified date, or a fabricated source.

The Rescue Prompt:

> Are these statistics and sources real, or are they placeholders? If you do not have live data access, state that clearly. Highlight any factual claims in your previous answer that I must verify, and point me to the official agency or website where I can find the actual numbers.

8. The Output Goes Off-Topic

The Diagnostic: The AI got distracted by a minor detail and went off on a frustrating, unrelated tangent.

The Rescue Prompt:

> Let's pause and reset. Go back to my original goal: [insert your original goal]. Ignore the last change or tangent, strip

away the unrelated details, and give me a revised answer focused only on that goal in a [insert format, e.g., bulleted list].

The Three-Strike Rule

If you use a targeted rescue prompt and the AI still fails to give you a usable output after three attempts, **do not keep fighting the chat window**.

Your current session may have too much extra information. Close the tab, open a new chat window to reset the AI's memory, and start again with a simpler, more organized prompt using the **Five-Part Formula**. Sometimes, starting fresh is the best way forward. If you want a better starting point, use the templates below.

Bonus 3: Privacy Safety Card

Treat any public AI chat box like a digital billboard. Once you paste information there, you cannot control where it goes. Many public AI models use your inputs to train future versions, so your private details could show up in someone else's answers.

Protect yourself, your family, and your workplace by using this **Privacy Safety Check** as a must-follow checklist before you hit "Send."

The "Never Paste" Checklist

Before you submit a prompt, ensure it contains **zero** of the following items:

- **Credentials & Access Keys**: Corporate or personal passwords, PINs, security questions, API keys, or Wi-Fi passwords.

- **Financial & Bank Details**: Credit card numbers, bank routing/account numbers, tax returns, invoices, and company financial ledgers.

- **Government IDs:** Social Security numbers, driver's licenses, passport details, and official tax IDs.

- **Medical Records**: Private patient history, specific diagnoses, prescription details, or physical and mental health reports.
- **Trade Secrets & IP**: Proprietary software code, unreleased product plans, internal business strategies, and copyrighted materials.
- **Real Names & Identifiers**: Full names, physical addresses, phone numbers, or email addresses of colleagues, clients, or family members.
- **Children's Data**: Names, ages, school names, schedules, or any identifying details of minors.
- **Legal Documents**: Active contracts, non-disclosure agreements (NDAs), court filings, and private dispute details.

The Golden Rule of AI Privacy

If you wouldn't write it on a postcard and mail it without an envelope, do not paste it into an AI tool.

The Safe Substitution Framework

You do not need to stop using AI for complex tasks. Just make sure to remove any personal details. Use the Safe Summary Method to reduce risk:

- **Instead of (Dangerous)**: "Karen Smith at ABC Marketing is underperforming and missed the Friday budget deadline."
- **Use (Safe)**: "Employee A at a mid-sized agency missed a project deadline. Draft a polite performance check-in email."
- **Instead of (Dangerous)**: "Review this diagnosis of asthma for my eight-year-old son, Leo."

- **Use (Safe)**: "What are standard, easy-to-understand questions a parent should ask a pediatrician about pediatric asthma management?"

The 3-Second "Pause & Check" Habit

Before you hit send, take a moment to ask yourself: "**Did I remove all real names, places, and sensitive numbers?**" Keep this safety card bookmarked, printed, or saved as a screenshot so you always stay in control.

Bonus 4: Everyday Prompt Templates

You can copy, adjust, and use these five main prompt templates right away. They follow the Five-Part Prompt Formula to help you get clear, natural, and useful results every time. Save them to your phone, notes app, or desktop for easy access.

1. The Professional Email Crafter

Goal: Create clear, balanced, and outcome-oriented workplace communication.

> Act as a professional communication coach. Help me write a clear, concise email about [insert situation, e.g., a project delay / a request for a meeting]. My goal is to [insert goal, e.g., establish a firm deadline / explain a mistake without sounding defensive]. The tone should be [insert tone, e.g., warm but firm/polite and collaborative/direct and professional]. Please keep the final message under [insert length, e.g., 150 words] and ensure it ends with a single, clear next step for the recipient. Give me three distinct subject line options.

Why it works: It forces the AI to respect length constraints and tone boundaries while preventing it from overcomplicating the call to action.

2. The Plain-English Tutor

Goal: Rapid, intuitive learning of complex or unfamiliar topics.

> Act as a patient, plain-English tutor. Explain the concept of [insert topic, e.g., compound interest/blockchain/inflation] to a complete beginner. Break down the concept using a simple, real-world analogy. Avoid technical jargon or academic definitions. After the explanation, list the top three common mistakes people make when trying to understand this topic, and then give me three multiple-choice questions to test my comprehension. Do not reveal the correct answers until I respond to the questions.

Why it works: It turns passive reading into an active, conversational learning loop, so you actually comprehend and retain the material.

3. The Interactive Project & Event Planner

Goal: Build highly customized, stress-free schedules and checklists.

> Act as an experienced, detail-oriented planning assistant. I need help planning [insert event/project/task, e.g., a two-week team rollout / a 3-day family road trip]. Before you generate any plan or schedule, please ask me exactly three clarifying questions about my constraints, budget, or preferences so you can tailor the plan to my needs. Once I answer, generate a chronological, step-by-step checklist divided into clear phases, and highlight the single most important task in each phase.

Why it works: By forcing the AI to interview you first, you bypass the "blank box" trap and ensure the final plan fits your actual constraints.

4. The Voice-Preserving Style Editor

Goal: Polish rough drafts while keeping your unique style intact.

> Act as a supportive, professional developmental editor. I am going to paste a rough draft below. Please improve this draft to be clearer, more engaging, and punchier, without changing my underlying meaning or making me sound robotic. Keep my authentic voice intact. Specifically, remove any passive voice, redundant sentences, and tired marketing clichés (like 'delve,' 'testament,' 'unlock,' or 'in today's fast-paced world'). After the revised draft, provide a brief bulleted list explaining exactly what you changed and why. Here is my draft: [paste rough text].

Why it works: It positions the AI as a collaborative editor rather than a hands-off ghostwriter, keeping you firmly in charge of your style.

5. The Skeptical Fact-Check Validator

Goal: Safeguard accuracy and verify high-risk claims before publishing or acting.

> Act as an objective, skeptical fact-checker. Analyze the text provided below. List every factual claim, date, statistic, quote, or historical assertion made in the text. For each claim, determine if it needs verification. If so, please tell me exactly which trusted, official primary source or website I should consult to double-check its accuracy. If you do not have live access to verify a claim yourself, say so clearly. Here is the text: [paste text or the AI's previous answer].

Why it works: It bypasses the AI's "confidence illusion" by forcing it to flag its own potential errors and point you toward real-world truth.

Bonus 5: Reader Self-Assessment

Real confidence does not come from memorizing complex code. It comes from seeing how much you have grown as a communicator. Before you start Chapter 1, take a few minutes to check your starting point. After you finish the last chapter and try the 7-Day AI Confidence Challenge, come back to this page to see your progress.

Rate your current confidence level for each statement on a scale of **1 (Low Confidence/Never Tried)** to **5 (High Confidence/Do This Easily)**.

The Human-in-Charge Scorecard

Key AI Skill & Competency Before Reading (1–5) After Reading (1–5)

1. **The Conversation Mindset:** I treat AI as an interactive, collaborative thinking partner rather than a transactional search engine box.
Before ______ After ______

2. **Structured Prompting:** I can easily build a powerful prompt using the five core pillars: Goal, Context, Role, Format, and Quality Check.
Before ______ After ______

3. **Privacy & Safety Boundaries:** I know exactly what sensitive data I must never paste into AI and how to anonymize my real-world details safely.
Before ______ After ______

4. **The Follow-Up Loop:** I can guide, adjust, and refine an initial weak AI draft using specific, constructive follow-up questions.
Before ______ After ______

5. **The Prompt Rescue System:** I can instantly diagnose why an AI output is bad (robotic, vague, or too long) and run a targeted rescue script to fix it.
Before ______ After ______

6. **Verification & Fact-Checking:** I can categorize AI tasks by risk level (Low, Medium, High) and systematically verify claims against trusted primary sources.
Before ______ After ______

7. **Voice & Style Preservation:** I can use AI as a writing assistant to break writer's block and organize ideas while retaining my unique, authentic human voice.
Before ______ After ______

8. **Repeatable Daily Workflows:** I know how to build simple, repeatable daily routines that integrate AI naturally into my planning, writing, and administrative tasks.
Before ______ After ______

TOTAL SCORE (Out of 40 Points)
Before ______/40 After ______/40

Understanding Your Score

Add up your points for both your "Before" and "After" scores to see your starting level and celebrate your progress:

- **8–15 Points (The Tech-Anxious Beginner)**: You may feel overwhelmed by the empty prompt box or worry about making mistakes. This is the perfect starting line! By shifting from a "search" to a "conversation" mindset, you will quickly find your footing.

- **16–30 Points (The Casual Experimenter)**: You have used AI for basic tasks but often get generic, robotic, or unpredictable results. This book will give you the precise troubleshooting tools and safety boundaries you need to take full control of your interactions.

- **31–40 Points (The Confident Human-in-Charge)**: You treat AI as a capable junior assistant, rescue bad answers systematically, and protect your data. You are running highly customized, safe daily workflows that save you hours.

Bookmark this page. Whether you feel "blank-box anxiety" like Paula, want to start a creative business like Sam, or are handling paperwork like Carol, you are about to learn a valuable new set of skills. Ready? Let's begin.

www.ingramcontent.com/pod-product-compliance
Lightning Source LLC
LaVergne TN
LVHW022217200926
848481LV00006B/2003